Frontispiece.

COLONEL ABBOTT, C.B.

HISTORY OF THE THIRTIETH LANCERS GORDON'S HORSE

MAJOR E. A. W. STOTHERD

The Naval & Military Press Ltd

Published by
The Naval & Military Press Ltd

PREFACE

THE material from which this history has been compiled was collected from the following sources:—The Manuscript Records of the Regiment; "Reports and Returns of the Hyderabad Contingent," by Colonel Hastings Fraser, Military Secretary to the Resident at Hyderabad, and formerly an officer of the 4th Cavalry; "History of the Hyderabad Contingent," by Major R. G. Burton; "The Revolt in Central India, 1857-59," also by Major Burton; "History of the 14th Hussars," by Colonel H. B. Hamilton; "History of the 13th Hussars," by C. R. B. Barrett; and lastly information kindly supplied by Mr. J. M. Bulloch, M.A.

The four Regiments of the Hyderabad Contingent Cavalry have been so closely allied and linked together in the past, that the history of one is often the history of the whole four, it being impossible to dissociate one from the remainder. To add to the connecting links, the British officers were constantly being transferred from one regiment to another. Hence a great deal of information about the other regiments will be found in this book.

In many despatches the Hyderabad Contingent Cavalry are comprehensively referred to as Hyderabad Cavalry, Nizam's Cavalry, etc., no numbers being given; in fact, they seemed to have been regarded, with considerable reason, as one large corps.

An effort has been made to preserve some of the brilliant record of the 3rd Lancers, Hyderabad Contingent, one squadron of which now forms part of the 30th Lancers (Gordon's Horse).

From a cavalry point of view, the most striking point appears to be the extraordinary mobility of the Hyderabad Contingent horsemen. Indeed, Lord Gough stated his opinion that they were the finest Irregular Cavalry in the world, and, a year or two later, Sir Hugh Rose described them as the wings of his operations.

E. A. S.

4th Dec., 1911.

CONTENTS

CHAPTER I.

The Origin and Evolution of the Hyderabad Contingent Cavalry.

PAGE

The Province of Berar—the Naiks—Pindarees—Mr. Russell's Reforms, 1811—the Reformed Horse, 1816—the Ellichpur Horse—the 1826 Reforms—the Raising of the 4th Nizam's Cavalry—Cavalry Pay—Arms—Horses—the 1854 Changes—Designation of Hyderabad Contingent introduced—General Wright's Reforms 1875—Changes in Horses, Transport, and Arms—the Hyderabad Contingent incorporated in the Regular Indian Army 1

CHAPTER II.

The Record of the Regiment.

The Raising of the Regiment—Early Services in the Field—Byam's March—Kurnool—Rai Mhow—Dharoor—Pursuit of Tantia Topee—Action at Chichambah—Rohillas defeated at Jintoor—Colonel Abbott—Record 1860-1911... 36

CHAPTER III.

The Indian Mutiny—Operations of the Malwa Field Force—The Campaign in Central India—The Gwalior Campaign.

Siege of Dhar—the Hyderabad Contingent Field Force—Capture of Piplia—Rawal—Battles near Mandesur—Storming of Goraria—Arrival of Sir Hugh Rose—Formation of the Central India Field Force—Capture of Rahatgarh—Saugor Relieved—Capture of Garhakota—Barole—Madanpur Pass—Siege of Jhansi—Battle of the Betwa—Storming of Jhansi—Kunch—Kalpi—Bilayan—Farewell Orders, etc.—Mutineers capture Gwalior—Actions at Morar and Gwalior—Re-capture of Gwalior—Pursuit and Action at Jaora-Alipur—Final orders—Casualties British Officers—Honours and Rewards 65

CHAPTER IV.

THE THIRD BURMESE WAR.

PAGE

Résumé of the Campaign—Condition of affairs August, 1886—Operations of the 3rd Cavalry—Pacification of the Ye-U District—Pursuit of Hla-U—Occupation of Wuntho—Pursuit of the Hlagaing Princes—Casualties 3rd Cavalry—Operations of the 4th Cavalry, 1st Squadron in Ye-U District—2nd and 3rd Squadrons in Kyouksai and Ava—Pursuit of Boh Tok—Boh Pyau Killed—Boh Shwe Yan Killed—the Sagaing District—Soobhan Khan's Successful Pursuit—Boh Ngou Killed—Return of 4th Cavalry to Mominabad—British Officers who served in the Campaign—Casualties—Honours and Rewards 144

CHAPTER V.

UNIFORM 167

CHAPTER VI.

SIR JOHN GORDON OF PARK AND THE GORDON REGIMENTS 173

APPENDICES.

(A) Officers who have Commanded the Regiment... 179

(B) Officers who have served in the Regiment (in Chronological Order) 185

(C) Officers serving in the Regiment on 1st January, 1911 198

Index 203

LIST OF PLATES

Colonel Abbott, C.B. ... ... *Frontispiece*

Facing Page

Hyderabad Contingent Cavalry. Indian Officer. 1846 16

Hyderabad Contingent Cavalry. Sowars. 1845 ... 24

Hyderabad Contingent Cavalry. Camel Sowar. 1846 32

Camel Sowar. 1902 40

Duffadar (Drill Order). 1897 48

Jat Sowar (Marching Order). Risaldar Major (Full Dress). British Officer. Sikh Duffadar. 1911 56

Sir Hugh Rose, K.C.B. (Baron Strathnairn) 80

Jhansi Fort from the South 88

Retribution Hill, Jhansi 104

Rough Sketch. Central India Campaign 142

Patrol in Burmese Jungle. 1888 146

A Troop at Ye-U. 1888 155

Hyderabad Contingent Cavalry. British Officer. 1845 168

British Officer. Drill Order. 1904 172

HISTORY

OF THE

THIRTIETH LANCERS

GORDON'S HORSE

CHAPTER I.

THE ORIGIN AND EVOLUTION OF THE HYDERABAD CONTINGENT CAVALRY.

THE Hyderabad Contingent Cavalry was originally raised from the armed levies of native chieftains and governors. It has been so often reorganised, and borne so many designations, that to trace it, and with it briefly also the artillery and infantry corps so long associated with it, step by step from the earliest period of its existence, it is necessary to go back to the history of the Berars. In this country it was raised, and with it assimilated, when every petty chief and nobleman had to maintain a number of armed followers for the protection of his life and property.

The whole of the Province of Berar was made over to the Nizam in 1804, at the conclusion of the

Mahratta War, "as a gratuitous concession to His Highness on the part of the British Government, and not surrendered to His Highness on the ground of his right to participate in the conquests effected by the war." As a matter of fact, the troops furnished by the Nizam in this campaign, some 6,000 cavalry and 2,300 infantry, do not appear to have done very creditably, and General Wellesley, in December, 1803, wrote to the Resident, suggesting the advisability of placing the Nizam's forces on a better footing. Owing to the opposition of His Highness, it was not till 1811 that any real progress was made.

The country had been overrun by Pindarees and Naiks to such an extent that the condition of the Berars, in 1811, was declared to have been declining for some years, both in population and in revenue. The Naiks were rebels and plunderers, but, except in their predatory habits, they differed not from other inhabitants of the province. They generally cultivated the land about their villages, but they neither paid rent nor acknowledged any authority, and they not only robbed travellers and plundered indiscriminately the whole country round them, but their power was so great, that on several occasions they compelled Government to enter into terms with them.

The villages of these Naiks were usually well fortified, and built in strong situations, protected by hills or jungles; the inhabitants were soldiers, armed with matchlocks and pikes; they generally fought on foot, some few Naiks had, however, horsemen and even Arabs employed under them.

There were five principal Naiks, who gave Government most trouble. These had each from 500 to 2,000 armed men, and property to the value of from half a lakh to three lakhs of rupees. The other less powerful Naiks were about twenty in number, dispersed over the country, each having 200 to 400 armed men, and from 8,000 to 25,000 rupees.

Pindarees were like the Mahrattas in their habits of life and warfare, but unlike them in not being united by nationality and one religion. From obscure freebooters, they rose into sufficient consequence to be deemed useful auxiliaries by different Mahratta powers, whose desultory mode of warfare was suited to their own habits. The object of the Pindaree was plunder, not war, and they were never known to fight when they could run away. They numbered from 8,000 to 10,000 horsemen, and had two battalions of infantry, and a foundry for casting cannon.

It was the existence of these freebooters which rendered the organisation of a disciplined force, which developed later into the Hyderabad Contingent, an imperative necessity. They traversed the Nizam's territory, and plundered towns and villages in all directions, and were a source of great anxiety to Government, which had to maintain a large force on the frontier to watch their movements. Owing to the ill-administration of the country, and the undisciplined state of the troops, the Naiks and Pindarees grew bolder and bolder, and their ranks were swelled by every Zemindar with a grievance against Government, real or imaginary, becoming themselves Naiks. The Nizam's troops, such as they were, had never been able to cope with these disturbers of the country with any extensive or lasting success. It was felt that, to afford peace and prosperity to the land, the power of the Naiks must be destroyed, and the only way to do it was to put the troops in a proper state of efficiency. This had been urged by the Resident as far back as 1804, who, as a preliminary measure, suggested the reorganisation of the cavalry on the Mysore *Silladar* system. However, as the Nizam refused to make any satisfactory arrangement for the regular payment of the men, the reorganisation had to be abandoned for the time being, but was not lost sight of. The difficulties to be overcome were the indifference of Government, and the jealousy of the Governors, Zemindars, and others, who did not like to see troops, hitherto dependent on them for their pay out of the *jaghirs* set apart for the purpose, quietly slip out of their hands.

The necessary reforms were gradually brought about, and in 1811, Mr. Russell, whose name will ever be associated with the Hyderabad Contingent, assumed the office of Resident at Hyderabad. He at once drew the attention of the British Government to the State of Berar, and pointed out that the Hyderabad Subsidiary Force could not, under the Treaty, properly be employed to restore order, and that the only way to effect this was to render the Nizam's military establishment efficient.

At this period the strength of the troops in Berar was 9,000 Cavalry, 5,000 Regular and 3,000 Irregular Infantry, and 25 guns. Of the Cavalry, 5,500 were *Sirkar* (regular), and 3,500 *Jaghirdar* (irregular). The

Regular Cavalry were considered the best in the Nizam's service, and fit for any duties required, but the Irregular Cavalry were only fit for the protection of their own *jaghirs*, and were scarcely employed for anything else. Both were on the *Silladar* system, and described as equal to the ordinary run of native cavalry in India. The Regulars received Rs.55 pay a month, and the Irregulars Rs.40, for horse and rider. Of the *Jaghirdar* cavalry, about 1,800 belonged to Salabut Khan, 900 to Soobhan Khan, and the remainder to different individuals, in small parties of 20 to 100 each.

One of the express conditions of the cavalry service was, that the horseman should receive no compensation for his horse if maimed or killed in battle, so, as his subsistence depended entirely on his horse, and as he generally had to borrow money to buy it, if he lost it, he could seldom raise money to buy another.

There were at this time six battalions of Regular Infantry, with two European officers attached to each. The men were armed, dressed and equipped like the Company's sepoys, and paid Rs.7 a month. Each battalion had two or more guns attached. The Irregular Infantry were only fit for police duty, except two battalions of Salabut Khan's, which were commanded by an Englishman named Drew.

Up to this period the Resident had no control over the Nizam's troops, and knew nothing more about them than he could learn from some of its officers. Lieutenant Sydenham did good service to the Nizam, in his endeavour to remodel, and place in a better state of efficiency, the troops of Berar; but he had scarcely left Berar when they drifted back into their old habits and customs.

Owing to one or two mutinies having broken out in the infantry, the Nizam's Government, urged on by Mr. Russell, at last, in 1812, decided on raising two battalions of regular infantry, equipped and disciplined like sepoys of the Company's Army. This scheme was carried out, and may be regarded as the first great step towards the formation of the Hyderabad Contingent. To these battalions were given the designation of the Russell Brigade, while to counterbalance the expense, some of the Irregular Infantry battalions were reduced, having been long since declared useless.

Up to 1815, no attempt at reform had been made in the cavalry branch of the Nizam's Army, owing to the peculiarity of this branch of the service, to its being entirely under native commanders, who had rights and privileges which could not be disturbed, and to the desire, in the then unsettled state of the country, to avoid the introduction of any measures that would tend to create suspicion and discontent among the troops, and the jealousy of the Rajas and others, who had control over these *risalas*.

It was with the greatest diffidence that the Resident, after having the papers under consideration for about three years, consented to urge the reorganisation of the cavalry branch. The necessity for this was now more apparent than ever, owing to the steady advance of the infantry in discipline and efficiency during late years.

The scheme for reorganisation was submitted April, 1816, and received the approval of the Government of India in July of that year. There were, however, many dissensions, and some delay caused by the obstructions raised by some Rajas, who perceived they were surrendering their authority over these troops, and being deprived of many perquisites derived from the method of paying them. Eventually it was agreed to reorganise in the Berars a body of 5,000 horse, of which 3,000 were to be *Sirkar* and 2,000 *Jaghirdar* troops. Another 1,000, required to bring the total up to 6,000, were to be supplied from Hyderabad.

Of the 3,000 *Sirkar* Horse, 2,000 were taken from men already in the service, and the remaining 1,000 were newly raised in place of men found unfit for the Reformed Horse. The *Jaghirdar* Cavalry was to consist of 1,500 of Salabut Khan's Horse, with 500 selected from other *Jaghirdars* in Berar, and to be stationed at Ellichpur. A party of horse, already organised on the Mysore *Silladar* system, were also to be raised in strength to 2,000, thus giving the Nizam a disciplined body of cavalry with a total strength of 8,000.

Of these various bodies of cavalry, the Berar or *Sirkar* Cavalry is the one which deserves special attention, as it was from it that the Hyderabad Contingent Cavalry eventually sprang. This force was now placed under the command of Captain Evan Davies, a Company's officer, who had recently been appointed to the Russell Brigade; he was assisted by five Company's officers, one of whom was to act as his Staff officer, and the remaining four to be employed with the *risalas* or divisions. Among these officers, Captain Davies became the first British officer of the 1st Nizam's Cavalry, Lieutenant H. B. Smith, 8th Madras Cavalry, of the 2nd Cavalry, and Lieutenant Wells, 7th Bombay Infantry, of the 3rd Cavalry. The cavalry were called the Reformed Horse, and divided into four divisions, each under command of a British officer, who drew pay from the Company, as well as from the Nizam. Of these divisions, three became eventually the 1st, 2nd, and 3rd Nizam's Cavalry; the 4th Cavalry never formed part of the Reformed Horse, but were raised later by Sir John Gordon, 13th Light Dragoons.

Captain Davies at once commenced entertaining new men and remodelling the old, and 2,000 were collected at Aurungabad, and 1,000 at Amraoti. The Corps soon became popular, and a few months later were ready to take the field in pursuit of Pindarees. Although each *risala* was under a European officer, yet, under ordinary circumstances, the men remained under command of their own native officers; the European officers held a general control over all, superintending establishments, and leading and directing the division when separately employed on service. The Reformed Horse were allowed three native commanders, with the pay of Rs.1,000 each per mensem; six native officers at Rs.500 each, and thirty at Rs.200 each. *Silladars* each received Rs.40 a month for self and horse, and the pay of *Bargirs* was Rs.15. Most of the men had firearms of sorts, the remainder were supplied on payment, from the Company's stores.

The Articles of War were not applicable, and the regulations issued for guidance on 15th April, 1816, were few and simple. 1st—Every man and horse was to be examined, approved, and registered by a European officer, or by some person delegated by him for that purpose. 2nd—All musters were to be made the same way. 3rd—The value of each horse was to be estimated and entered in a register, and the owner of any horse killed or disabled on service was to be paid the value by Government. 4th—The European commanding officer was to see that every man was regularly paid, and to receive and redress all complaints, which might be made him of injustices and severity of any kind. 5th—*Silladars* and *Bargirs* should receive their pay without any deductions or stoppages whatever.

The Reformed Horse never had any European non-commissioned officers attached to them, and only one British officer per regiment. They were continually in the field, and did excellent service against the Naiks and Pindarees, and, besides numerous smaller actions, were present at the battle of Mehidpur and the siege of Nowah. From 1817-1825 they were continuously employed reducing forts, and breaking up bands of rebels and marauders. Their numbers at this time were 4,000, while the *Jaghirdar* Cavalry were 2,000 strong, of which Salabut Khan had 1,500 horse, whose duty was to defend the territory near Ellichpur.

It is interesting to note that, about 1816, in order to make Russell's Brigade complete in itself, a regiment was raised called the "Russell Cavalry." It was on the regular system, had horses and equipment supplied by Government, and a number of men of the Madras Cavalry attached to it. The establishment was 5 British officers, and an assistant surgeon, quartermaster, sergeant-major, quartermaster-sergeant, 2 corporals, 294 natives of all ranks, 300 horses, *salootris*, nalbunds, grass-cutters, etc. On the score of expense, this regiment was broken up on the 21st December, 1821; the horses were at once sold, and the men were allowed to continue a short time as dismounted cavalry, when they were either discharged or drafted into other corps.

At the close of 1819 it was determined to reduce the numbers of the Reformed Horse; so 429 men of the Hyderabad *risala* were at once discharged, and the remainder of the Corps were brought to Hyderabad, where it was allowed to die out gradually as casualties occurred. Thus only three *risalas* remained till 1826, when the 4th Cavalry was raised.

In April 1821, an additional British officer, with an allowance of Rs.500 per mensem, was sanctioned for each of three cavalry corps. It had been found that only one British officer per regiment was not sufficient, owing to some being always absent on leave or sick.

During December, 1821, the Headquarters of the Reformed Horse was established at Mominabad, to which place Captains Sutherland and Smith were directed to march their *risalas*. At the same time the number of troopers in each *risala* was reduced to 700 men.

Of the *Jaghirdar* Cavalry, Salabut Khan's Horse, now known as the Ellichpur Horse, remained at

Ellichpur, and consisted of 2 British officers, 583 natives of all ranks, and 17 followers. Frequent complaints were made against Salabut Khan for neglecting to pay his troops at the proper time. He had certain districts assigned to him for the pay of the Ellichpur Brigade.

The year 1826 may be looked upon as the most important period in the organisation of the Hyderabad Contingent. The troops had hitherto been considered in the strictest sense local, for not only were they never moved out of their respective brigades and divisions, except for active service, but they had the designation of the brigade and division in which they were serving, each being separate and distinct, and governed by its own local rules. In the absence of general orders, which had hitherto not been in use, the orders of the Resident were communicated to officers either by letter or memorandum.

All this was now changed; the Corps were formed into one Army, and regiments and batteries numbered throughout, without reference to the brigade or division in which they were serving. General orders were issued for the first time in June, 1826, and, at the same time, a printed code of "Rules and Regulations for the Guidance of the Nizam's Army" was issued, under the authority of the Resident at Hyderabad, on the part of the Nizam's Government.

The object of the reorganisation of 1826 was to equalise the strength of various regiments, place them all on the same footing with regard to pay, etc., and reduce the expenditure of the military establishment without interfering with efficiency. Thus it came about that the 4th Nizam's Cavalry, now the 30th Lancers, was then raised.

At this time the Nizam's Regular Army, viz., that commanded by European officers, consisted of two divisions, two brigades, and one brigade of cavalry, which were distributed at the following principal stations.

The 1st, or Hyderabad Division, had its headquarters at Bolarum, and consisted of two regiments of infantry, a corps of engineers, and a company of artillery, with the usual proportion of gun lascars and ordnance drivers. A detachment of five companies of infantry, under a European officer, was furnished to Mahadapur on the south bank of the Godavery, for the purpose of curbing the rebellious zemindars in that direction, and another company was employed in the Medduk district. This division had also attached to it a squadron of Reformed Horse, a battering train, and store department.

The 2nd, or Aurungabad Division, consisted of two battalions of infantry, a company of artillery, with battering train and store department, also two irregular battalions, called respectively the garrison and invalid, the former protecting the north-west *ghauts* against incursions of Bhils, the latter performing the duties of police. Numerous detachments from this division were stationed at convenient distances between Ajunta and other *ghauts*, to guard the frontier. The Headquarters of a cavalry regiment were also stationed at Aurungabad, but the greater part of this regiment was detached in the districts, wherever their services were required.

The Hingoli Brigade originally formed part of the Aurungabad Division, but in January, 1825, it was made into a separate and independent command of two regiments of infantry, one company of artillery with four light guns, a detachment of cavalry, and a store department.

The Ellichpur Brigade consisted of one cavalry regiment, one battery of artillery with horses, two battalions of infantry, and a store department. It was maintained and paid under special engagements with the *Jaghirdar*, Nawab Namdar Khan; its situation on the frontier rendering it a most important post of the Nizam's army.

The Headquarters of the Cavalry Brigade was at Mominabad, a very healthy station, and in those days strategically important as being central and guarding various important *ghauts*. For cavalry it had the inestimable advantage of a most excellent grass supply. Major Evan Davies was still in command of this brigade at the time of the 1826 reorganisation, and in the following report, dated March, 1826, gives a good idea of the state of the cavalry at the time.

He writes:—"In 1816, I was ordered to Aurungabad to reform a part of the Berar Horse. Three thousand of the Hindustani quota were given over to me by Captain Sydenham. I found the men of the best description, mostly Mohammedans from the north of Hindustan, mixed with a few foreigners from Baluchistan and Sind. The whole were, however, badly mounted, and without the least pretence to discipline of any kind.

"They were formed into three *risalas* or corps, divided into ten troops, each troop commanded by a jemadar. One European officer superintended the *risala*, assisted by a native commandant and two risaldars.

"The unserviceable horses were replaced by good ones, and each was valued and registered by the European officer.

"The matchlock was laid aside, as being of no use on horseback; on the contrary it invariably discovered our march to the enemy at night. Employed as we were, without infantry, and being frequently obliged to act on foot, I procured English carbines for one-third of the men; the remainder I armed with a pistol, sword, and spear each. As, however, they had always been in the habit, under their own chiefs, of marching in one long, loose, extended line, it was absolutely necessary to teach them to wheel by threes, and to form line in any given direction. This they soon acquired.

"In November the same year, I was joined by one thousand five hundred men from Hyderabad, consisting for the most part of Pathans and Moguls. These were ordered into the Nagpore territories, under the command of Captain Pedler, and disbanded shortly after the war.

"The country being overrun with banditti, obliged the Government to detach the horse in small parties for its protection, to command which we were obliged to promote four duffadars, or non-commissioned officers, to each troop.

"The *risalas* were soon reduced to seven hundred privates, and the senior risaldars retired on pension, leaving the Brigade composed of three *risalas*, each having a captain commandant, ten jemadars, twenty duffadars, and six hundred and eight privates. Two-thirds of the brigade is detached over the country, the remainder are at Mominabad, and are exercised in the following manner. The carabiniers are taught to skirmish on foot and horseback, and are excellent marksmen. They are also taught the use of the sword, agreeably to the native mode. The spearmen are taught the native spear exercise, and to skirmish on horseback.

"The *risalas* are told off into squadrons for field exercise; and are merely taught to change position on a flank by bringing forward or throwing back a wing; form close column of squadrons and form line from this column; attack to the front both by squadrons and in line; retire in line by alternate half-squadrons. The above is all they have attempted, as more attention has been paid to their using their arms well, singly as skirmishers, than to making them regular troops.

"The men composing the brigade are generally of the same description as those afore-mentioned, except a few lately enlisted in His Highness's dominions. They are mounted on the Deccani horse, rather under the standard of the regular cavalry, but capable of undergoing great fatigue, and subsisting on very little grain."

In 1826, when the Nizam's Army was reorganised, and the different corps received new titles, the cavalry had been considerably reduced in numbers; but a new regiment, the 4th Cavalry (now the 30th Lancers), was raised and commanded by Captain Sir John Gordon, Bart., who had shortly before been seconded for service with the Nizam's Army from the 13th Light Dragoons, and given command of the Ellichpur Horse.

The establishment of each of the four cavalry corps was now fixed at two European officers, a native commandant, or risaldar, 8 jemadars, 16 duffadars, 16 naib duffadars, and 512 troopers.

It may here be noted, that these designations did not correspond with those of native ranks in the Presidency armies, the risaldar being the same as the risaldar-major elsewhere, while the duffadars were native officers in the Nizam's Cavalry. These anomalies, however, continued till 1856, when orders were issued that in future risaldars were to be styled risaldar majors, jemadars - ressaidars, duffadars-jemadars, and naib duffadars-duffadars.

In 1816, when Captain Davies assumed command of the Reformed Horse, there were four native chiefs of *risalas*, all of whom were retired in 1825. The most famous of these was Nawab Murtaza Yar Jung, whose *risala* in 1826 became the 2nd Cavalry; he was as much distinguished for his magnificent fighting record and bravery, as for his loyalty and friendship for the British.

From the foregoing descriptions and statistics, a pretty accurate idea may be obtained of the composition, training, equipment, etc., in 1826, of the regi-

Facing Page 16.

HYDERABAD CONTINGENT CAVALRY,
INDIAN OFFICER.
1846.

ments of the Nizam's Cavalry, including the newly raised 4th Cavalry.

All regiments and batteries were now for the first time numbered. The cavalry were as follows:—

(1) Nawab Jalal-ul-daulah's, Captains Davies' and Clerk's *risala.* } became in 1826 1st Nizam's Cavalry.

Subsequently—

in 1854, 1st Cavalry Hyderabad Contingent.
in 1890, 1st Lancers Hyderabad Contingent.
and in 1903, 20th Deccan Horse.

(2) Nawab Murtaza Yar Jung's, Captains Hallis' and Smith's *risala.* } became in 1826 2nd Nizam's Cavalry.

Subsequently—

in 1854, 2nd Cavalry Hyderabad Contingent.
in 1890, 2nd Lancers Hyderabad Contingent.
and in 1903, 29th Lancers (Deccan Horse).

(3) Rai Barcha Mull's, Captain Wells' *risala* } became in 1826 3rd Nizam's Cavalry.

Subsequently—

in 1854, 3rd Cavalry Hyderabad Contingent.
in 1890, 3rd Lancers Hyderabad Contingent.
and in 1903 was divided up, one squadron being sent to each of the 1st, 2nd, and 4th Regiments.

(4) The regiment newly raised by Sir John Gordon. } became in 1826 4th Nizam's Cavalry.

Subsequently—

in 1854, 4th Cavalry Hyderabad Contingent.
in 1890, 4th Lancers Hyderabad Contingent.
and in 1903, 30th Lancers (Gordon's Horse).

(5) Salabut Khan's *risala*, the Ellichpur Horse } became in 1826 5th Nizam's Cavalry.

The last regiment was, as explained above, always distinct in every respect from the other four. It was disbanded in 1854. The horses were supplied by Government, similar to the regular cavalry of the Madras Army.

There were four companies of artillery, numbered 1 to 4, distributed at Bolarum, Aurungabad, Hingoli, and Ellichpur.

The infantry regiments were numbered 1 to 8, and of these the first and second regiments were the first and second battalions Russell's Brigade respectively. The seventh and eighth regiments were the two regiments of the Ellichpur Brigade.

The pay of the cavalry regiments was fixed as follows:—

British Commandant ...	Rs.1,000 per mensem.
British Lieutenant	Rs.1,000.
Assistant Surgeon	Rs.600.
Risaldar (Native Commandant)	Rs.500.
Jemadars	Rs.200.
Duffadars	Rs.60.
2nd Duffadars	Rs.45.
Horsemen	Rs.40.

A full establishment of followers was allowed at fixed pays, and an office allowance of Rs.100 a month to the adjutant, and Rs.100 for the purchase of country medicines.

The total cost of each regiment per mensem was Rs.27,108.

From the 1st January, 1828, the establishment of the Cavalry Brigade was fixed as follows for each regiment:—

			Rs.		
1	Risaldar	at	500	a	month.
8	Jemadars	"	200	"	"
24	Duffadars	"	60	"	"
24	Naib Duffadars	"	50	"	"
1	Trumpet-Major ...	"	50	"	"
8	Trumpeters	"	40	"	"
480	Horsemen	"	40	"	"
1	Risaldar Mutsaddi ...	"	50	"	"
1	Camel Nagara	"	20	"	"
4	Camel Harkaras ...	"	30	"	"
8	Harkaras	"	7	"	"
1	Head Bearer	"	9	"	"
5	Bearers	"	8	"	"
1	Armourer	"	15	"	"
1	Bellows-boy	"	6	"	"
8	Troop Mutsaddis ...	"	20	"	"

Thus the total cost of the native ranks and followers was now Rs.24,786 a month. British officers and assistant surgeon would probably make this Rs.2,600 more, making the total cost approximately the same as in 1826, but the establishment was different.

In June, 1828, the Cavalry Division consisted of sixteen squadrons (four per regiment), made up of 204 *Silladars*, exclusive of officers, and 1,716 *Bargirs;* of these, 1,680 were Mohammedans, 196 Rajputs, 36 Sikhs, and 8 Mahrattas.

The following extract from a report by Major-General Sleigh, C.B., who inspected the division in March and April, 1832, is of interest:—

"The horses of the division are chiefly bred on the banks of the Bheema river and the adjacent country; they generally show a good deal of blood; the majority are good, useful horses, standing fourteen hands two inches; *they make extraordinary marches of sixty miles in a night, and seldom or ever leave a man behind*; *indeed, the men would fancy themselves disgraced, if unable to proceed with their comrades.* They are repeatedly called out, and no trouble attends their moving at the shortest notice. Generally speaking, they are in very good condition, and are without doubt a most useful and effective body of horse to the State.

"The average duration of a horse is ten years, and the cost, I should imagine, to be from two hundred and fifty to three hundred rupees. The weight of the men, including everything they carry, is rather above eleven stone.

"The great advantage these corps possess over the Regular Cavalry is, they move without delay, and never require aid or assistance from the commissariat; they have everything within themselves that the most efficient commissariat could give a King's regiment, and could, on emergency, get twenty-four hours' start of any of them."

Although the 5th Nizam's Cavalry, or Ellichpur Horse, both in origin and composition, was totally different from the four regiments, which became the Hyderabad Contingent Cavalry, and indeed was always kept quite distinct from them, it is noteworthy that, on 24th July, 1828, its composition and cost was as follows:—1 Commandant, 1 lieutenant, and 1 assistant-surgeon (British), 600 natives of all ranks, including dhooly bearers and dressers, and 579 horses supplied by Government, as in the Madras Cavalry. The total annual cost of the regiment amounted to Rs. 3,38,328, or Rs.28,194 a month; slightly more than the four irregular regiments.

Prior to the year 1841, the infantry and artillery were governed by the Articles of War then in force in the Madras Army, whilst offences in the cavalry branch were tried by native assessors, the commission being known as a *panchayat*. A new code was promulgated to the Contingent in July, 1841, with the express sanction of the Governor-General of India in Council. The necessity was then urged of making the same code applicable to the Contingent, as was applicable to the Army in India. A General Order, dated 10th March, 1848, was accordingly approved by the Governor-General, making the Articles of War for the Native Army in India applicable to the Contingent. The cavalry were, on the recommendation of the Resident, exempted from the operation of this order, and retained the system of trial by *panchayat* until March, 1856, when they were made amenable to the Articles of War. It was not until

17th December, 1897, that by a notification of the Government of India, the provisions of the Indian Articles of War of 1869, as amended by the Act of 1894, were made applicable to the Hyderabad Contingent.

On 4th March, 1843, the 5th Nizam's Cavalry, which had hitherto been separate and distinct from the other four regiments, was added to the Cavalry Brigade. The strength of each regiment was revised as follows:—

			Rs.	
1	Risaldar	... ... at	500	a month.
8	Jemadars	... ... „	200	each „
16	Duffadars	... ... „	60	„ „
64	Naib Duffadars	... ... „	50	„ „
1	Trumpet-Major	... ... „	50	a month.
6	Trumpeters	... ... „	40	each „
2	Kettle-drummers	... „	40	„ „
480	Troopers	... ... „	40	„ „
4	Camel Gunners	... ... „	30	„ „
1	Regimental Mutsaddi	„	50	a month.
8	Troop Mutsaddis	... „	20	each „
7	Harkaras	... ... „	8	„ „
1	First Dresser	... ... „	105	a month.
1	Second Dresser	... ... „	76	„ „
1	Native Dresser	... ... „	36	„ „
2	Hospital Servants	... „	8	each „
1	Schoolmaster	... ... „	15	a month.
1	Armourer	... ... „	15	„ „
1	Bellows-boy	... ... „	6	„ „
8	Bearers	... ... „	5	each „
1	Bazaar Kotwal	... ... „	30	a month.

The authorised number of British officers at this period was four, including the surgeon. In 1843, a return shows the total strength of the five regiments to be:—

British officers 22, warrant and non-commissioned officers (probably British hospital dressers) 10, native ranks 2,910, followers 180.

It was estimated that each regiment now cost Rs.26,604 a month.

As above mentioned, when Captain Davies assumed charge of the Reformed Horse in 1816, the matchlock was discarded, English carbines procured for one-third of the men, and the remainder armed with pistol, sword, and spear.

An order, dated 1st May, 1842, directs that half the troopers shall have lances; but this was cancelled in February, 1848, when orders were issued that "as many as wish to have carbines instead of lances are to be allowed the former." From 1843 percussion firearms were gradually introduced in place of the flint-locks, and remained in use till the adoption of the Victoria carbine in 1860.

In 1845, the cavalry were generally mounted on Arabs and country breds. In this year the classes of horses in the five regiments were:—

Arabs	358	Heratis	1
Deccanis	1,554	Kathiawars	4
Hindustanis	298	Khelatis	1
Wilaitis	79	Not known	375
Persians	58	Wanting to complete	7

The average weight carried by the horses of the Nizam's Cavalry, at this time, was 13½ stone. Being light, and consequently mobile, these regiments were able to start at an hour's notice, ride a hundred miles straight away, fight an action at the end of it, and return at once to their cantonments.

From 1st January, 1854, the designation of the force was changed to "Hyderabad Contingent," and the following innovations made in its constitution: The Hingoli, Ellichpur, and Cavalry Brigade Commands (the several brigadiers of the Nizam's Army held semi-political charge of their frontier posts, and reported direct to the Resident such occurrences as came under their notice) were abolished, and the force placed under the command of two brigadiers, whose commands were styled the Northern and Southern Divisions. Each had a staff officer for the performance of the duties of brigade-major and paymaster. The 5th Cavalry, also the 5th and 6th Infantry were broken up, the 7th and 8th Infantry being subsequently known as the 5th and 6th.

The four batteries of artillery were now each commanded by a Royal Artillery officer, and had six field guns, (four six-pounders, and two twelve-pounder howitzers), drawn by bullocks.

The cavalry were now to consist of four regiments, of 500 troopers each, in three squadrons.

The strength of each of the six infantry regiments was raised to 800 privates.

The total strength of the four cavalry regiments was at this time: British officers 16, British warrant and non-commissioned officers 10, Indians 2,300, followers 96.

Facing Page 24.

HYDERABAD CONTINGENT CAVALRY.
SOWARS.
1845.

The order concludes:—"The Hyderabad Contingent will consist of not less than four field batteries of artillery, 2,000 cavalry, and 5,000 infantry."

The Hyderabad Contingent Cavalry had now established their reputation for mobility, efficiency, and fighting qualities; and in 1853 Lord Gough, lately Commander-in-Chief in India, stated before a committee of the House of Commons his opinion, that they were the finest Irregular Cavalry in the world. This estimate is borne out by the historical record of their deeds. Four years later, in the Central India Campaign, under Sir Hugh Rose, they amply proved the truth of Lord Gough's statement. The reason for their efficiency is not far to seek. Burton, in his "History of the Hyderabad Contingent," writes:—"The history of the force has now been brought up to the dark period of the great Mutiny, in the suppression of which the Hyderabad Contingent was about to play so distinguished a part. They were indeed well fitted for action. For forty years they had been engaged in continual fighting, and scarcely a month had passed without some portion of the force being on active service in the field, so that, in fact, they may be said to have been a field force from the very commencement. They passed as much time on active service as they did in their small and isolated cantonments, and their training was consequently such as to fit them best for undertaking their rôle in a larger theatre of war."

"In view of the paucity of officers with the Hyderabad Contingent, the efficiency of the corps is remarkable. . . . . Indeed, the paucity of British officers has the advantage, that it throws more work, responsibility, and independence upon the native officers, thus increasing their efficiency, which is so liable to deteriorate when everything is done by the British officers. Perhaps it is to this fact that the extraordinary efficiency of the Hyderabad Contingent Cavalry is to be mainly ascribed."

In 1855, the Hyderabad Contingent Cavalry were made eligible for pensions, at the rates for the same branch of the Bengal Army; hitherto there had been no pensions for them.

The staff pay of British officers was readjusted in 1863. Each officer received pay of rank and in addition the following staff pay:—Commandant Rs.700 a month, second-in-command Rs.300, adjutant Rs.250, and other officers Rs.150.

In 1864, compensation for dearness of provisions was made applicable to the Hyderabad Contingent. The batteries were reduced to four guns drawn by horses, and a subaltern officer took the place of the warrant quartermaster, the European staff-sergeants being subsequently abolished. In December of the same year the command of the Hyderabad Contingent was declared to be a 1st class brigade, and the officer commanding received the allowance of a 1st class brigadier, in addition to staff corps pay of rank.

In 1875, Brigadier-General T. Wright, C.B., who had been appointed to the command of the Hyderabad Contingent from the 11th Bengal Lancers, set about remodelling the cavalry *Silladari* system on the lines of the Bengal Cavalry. The system hitherto prevailing in the Contingent Cavalry was known as the *Pagah* system, under which any man, whether belonging to the regiment or not, could be the owner of two or more horses (*Assamis*). Two or more horses formed a *Pagah*, the owner was a *Silladar*, and the ownership of each horse an *Assami*, whilst the man who rode the horse, unless himself the owner, was a *Bargir*.

When the cavalry was reformed, and until the reorganisation now under review was introduced, many people—chiefs, traders, and others—owned large *Pagahs*, in some cases as many as a hundred horses or more.

In 1860, however, orders were issued that *Assamis* could be owned in future only by those actually enlisted in the regiments. At the time when *Assamis* could be owned by anyone, irrespective of whether they were in the regiment or not, their price was very high, amounting to as much as *H.S. Rs.1,000 or more, but, when the ownership was restricted to enlisted men, they fell off considerably in value.

When General Wright carried out the reforms necessary to bring the Hyderabad Contingent on the same footing as the Regular Presidential Armies, steps were immediately taken, under the orders of Government, to transform gradually all four cavalry regiments into *Khodaspahs*, or one-horse *Silladars*, so that each man should own the horse he rode.

The price of an *Assami* was fixed at H.S. Rupees 450, including the horse, saddlery, horse clothing, etc., as well as a half share in a pony.

---

* The coinage in use in the Nizam's Dominions was termed *Halli Sicca*, the exchange with Government Rupees varied considerably.

By the orders of Government, every *Silladar*, on being pensioned, was obliged to sell his *Assamis* to the *Bargirs* of his regiment, for the sum of H.S. Rupees 450. In this manner all the *Bargirs* gradually became *Khodaspahs*, each man owning his own horse and equipment. It took about fifteen years (in some regiments more) to bring about this reform. As few men were in possession of sufficient funds to purchase *Assamis*, advances were made from the regimental cash chest by the commanding officer to assist them. An advance of Rs.10,000 was made to each cavalry regiment by Government, to be loaned without interest to deserving *Bargirs* to enable them to purchase horse *Assamis*. These loans were recovered by monthly instalments.

Many other improvements were also introduced this year. The Victoria carbines, with which the regiments had been armed since 1860, had been replaced, in January, 1871, with new-pattern, smooth-bore carbines and pistols. Now, in 1875, regular musketry training, equitation, and foot-drill were introduced, and all the men were obliged to attend at morning and evening stables, and groom their own horses, which hitherto they had never done. All regiments were equipped with saddlery, resembling in pattern that of the British cavalry, in lieu of the native saddlery (*khogeer*), which they had always used up to this time. Other equipment was gradually brought up to the pattern and standard of more regular corps. British officers were placed in charge of squadrons, and the appointment of "Woordi Major" was now instituted. Finally the first camp of exercise on record, that the Hyderabad Contingent Cavalry attended, was held by General Wright at Jalna in January, 1875. Till then their training had been entirely in the field on the active service, upon which they were perpetually employed.

With a view to improve the condition of native commissioned and non-commissioned officers and men of the Hyderabad Contingent, the Governor-General in Council was pleased to sanction the number of troopers in each cavalry regiment being reduced to 478, to provide for increased rates of pay to all native ranks from 1st January, 1877. The strength of regiments was accordingly reduced by 22 troopers.

The pay of commissioned officers was increased as follows:—

			Rs.		
1	Ressaidar	... ... at	200	per	mensem.
1	"	... ... "	185	"	"
1	"	... ... "	170	"	"
3	"	... ... "	150	"	"
1	Woordi Major	... "	120	"	"
2	Jemadars	... ... "	80	"	"
2	"	... ... "	70	"	"
2	"	... ... "	60	"	"

The risaldar-major appears to have remained at Rs.300 per mensem.

In place of six jemadars (to be absorbed), six kote-duffadars were to be appointed at Rs.49 per mensem.

The pay of sowars remained at Rs.30. Sowars of the Presidential Armies at this time only received Rs.27, but the Contingent Cavalry received no good conduct pay. This arrangement was found, however, advantageous for recruiting in Northern India.

A farrier-major and salootri were also sanctioned on the same pay as duffadars, viz., Rs.41-8 per mensem; a quartermaster duffadar with allowance of Rs.9; six pay duffadars with Rs.6; and two lance-duffadars per troop, each with Rs.2 per mensem. Six troop *Mutsaddis* and four *Harkaras* were dispensed with.

Forge funds and workshops were now established. Monthly subscriptions to the *Chundah* fund were doubled, and the fund paid the full price of remounts.

In 1827 the lance in use was nine feet long, with a pennon thirty inches long and twelve inches broad.

During 1878 the 4th Cavalry were entirely armed with lances, each man also having a carbine and a sword. The lance was still nine feet long, with a hog-spear head of bay leaf pattern; in fact, all the heads were manufactured by the leading hog-spear makers of the day; the pennons were red and white. By 1880, the entire Hyderabad Contingent Cavalry were armed with the lance. About 1903 the lances were reduced in length to eight feet six inches.

From 1st July, 1882, good conduct pay was granted to non-commissioned officers of the Hyderabad Contingent, at the rate of Rs.1 after two years' good service, Rs.2 after four years, Rs.3 after six years, and Rs.4 after eight years.

On 1st April, 1883, the tenure of commands of the Native Army, including the Hyderabad Contingent, was limited to seven years.

During 1884, Snider carbines and revolvers were received in place of the smooth-bore carbines.

In 1888 an additional squadron officer to each regiment was sanctioned, and the following year another; thus bringing up the number of British officers to seven, including the medical officer. During the Burmese Campaign, the want of more officers was clearly shown by the fact, that it was found necessary to deplete the remaining regiments of their British officers, in order to provide officers for the corps proceeding on service. The number was further raised to eight per regiment in 1892 and 1893, whilst at the same time squadron commanders were appointed on the same rates of pay as in the Bengal Army. In ensuing years, the Hyderabad Contingent was kept on the same footing as Bengal as regards the supply of British officers.

In 1894, the full complement of Martini-Henry carbines was received by all regiments, and the Sniders were returned to the arsenal.

It was about this period, that various changes were beginning gradually to make themselves felt in all the Contingent Cavalry regiments. During the later years they had been mounted to a great extent on Arabs, purchased in the Bombay Market, with a few Persians and country-breds. The supply of Arabs was now beginning to decrease, while a much improved stamp of small Australian horses began to be imported into India. Thus it came about, that the Australian some ten years later supplanted the Arab as the main source of supply of remounts.

Again, till after the Burmese War, ponies had done all the grass and transport work of the regiments. Mules then gradually took their place, till the ponies were entirely eliminated.

Lastly, saddlery of the most up-to-date pattern was adopted, in place of the rather heavy, cumbersome saddles, with high brass-bound cantels and cruppers and breast-plates, that had been in use since 1875.

From 1st April, 1895, an increase of pay of one rupee per mensem was granted to sowars, and they also became eligible for good conduct pay. This placed them on exactly the same footing as the rest of the Indian Army. The numerical strength of each regiment was reduced to 500 natives of all ranks. The class squadron system was now introduced, whilst the Force, which had hitherto been solely under the orders of the Government of India, was placed directly under the Commander-in-Chief for the regulation of its training.

In 1901, Lee-Enfield carbines were issued to replace the Martini-Henry. These were only used for about two years, and in 1903 were discarded in favour of the Lee-Enfield rifle.

In 1902, a fresh agreement was concluded with the Nizam's Government, under operation of which Berar was leased in perpetuity to the British Government, and the Hyderabad Contingent was, in April 1903, incorporated with the Regular Indian Army. The four cavalry stations at the time were:—Bolarum, with a garrison of one cavalry regiment, one battery, and one infantry regiment; Aurungabad and Hingoli, with similar garrisons (except that, at one time, a detachment of a squadron of cavalry used to be furnished to Ellichpur from Hingoli); and Mominabad, an isolated place, eighty miles off the line of rail, which had one cavalry regiment. Of all the old

Facing Page 32.

HYDERABAD CONTINGENT CAVALRY.
CAMEL SOWAR.
1846.

cantonments, only Aurungabad and Bolarum were retained. The cavalry were transferred at first to the Bombay Command, till Lord Kitchener's reorganisation brought the whole of the Indian cavalry together. It was decided to make of the Hyderabad Contingent Cavalry three regiments of four squadrons; as each of the four old regiments had only three squadrons, this was effected by transferring one squadron of the 3rd Lancers to each of the other regiments, which now became:—

1st Lancers Hyderabad Contingent—20th Deccan Horse. Composition, one squadron Sikhs, one squadron Dekkani Mohammedans, one squadron Hindustani Mohammedans, one squadron Rajputs or Jats.

2nd Lancers Hyderabad Contingent — 29th Lancers (Deccan Horse). Composition, one squadron Sikhs, one squadron Dekkani Mohammedans, two squadrons Jats.

4th Lancers Hyderabad Contingent—30th Lancers (Gordon's Horse). Composition, two squadrons Sikhs, one squadron Northern Mohammedans, one squadron Jats. One of the Sikh squadrons was transferred from the 3rd Lancers.

The Mohammedans had long remained the nucleus of the cavalry, but, beginning from the eighties, Sikhs, Jats, and a few Rajputs had been introduced, and, as has already been mentioned, the class squadron system was regularly adopted in each regiment in 1895.

This brings the history of the Hyderabad Contingent Cavalry to an end. For the whole course of their service, the four regiments had been closely associated and linked together. Many of the British officers had served in two or three of the regiments, and some even in all four, in fact, both officers and men were frequently transferred. The native ranks generally had relations serving in the other regiments.

It is pleasing to notice, that the officers of the three remaining regiments are now permanent honorary members of each other's messes.

Burton, in his "History of the Hyderabad Contingent," sums up its services as follows:—"It is only the student of history who, in the course of his researches, becomes aware of the progress of events which have led to the pacification of India, and is able to estimate at their proper value those great deeds that won and kept the Empire. In those events and those deeds, the Hyderabad Contingent has played a not inconspicuous part. Its corps participated in the victories of the Mahratta War, 1817; they contributed largely to the pacification of the country during the ensuing forty years; and they formed a great factor in the suppression of the Mutiny in Central India in 1857-8, and in the maintenance of peace in Southern India during that dangerous period. The story of their deeds, which are mostly those of a past generation, has long been buried in the archives that lay beneath the dust of forty years. Now that these have seen the light, it is to be hoped that the Force will receive its just due in the history of the conquest and pacification of India, and that, though dead, it will not be forgotten.

"The deeds of Evan Davies, of George Hampton, of John Sutherland, of Murtaza Yar Jung, and

*Ahmad Baksh Khan, who rode forth so often through the Deccan at the head of their gallant horsemen, who made great marches, and performed great acts of valour, illuminate the page of history, and lend a spirit of romance to the stories of the campaigns, in which they bore so conspicuous a part. Our native army is justly proud of its records, and no portion of it has a greater right to such pride than the squadrons, battalions, and batteries of the Hyderabad Contingent.

* * * * * * * *

"When the Indian Mutiny broke out, the Hyderabad Contingent not only remained true to its salt, but by its attitude and its deeds saved the situation in Southern India. And when a large portion of the Force took the field in Central India, they earned for themselves a great reputation by deeds which are unsurpassed by those of any of our native troops. It was not in vain that Sir Hugh Rose, after they had marched a thousand miles with him and distinguished themselves in many actions, called them 'the wings of the army.' Their record in the Mutiny stands for all time emblazoned on the pages of the history of that period, and the names of Abbott, of the Orrs, of Dowker, and Clogstoun, will not readily be forgotten."

* 3rd Cavalry.

CHAPTER II.

The Record of the Regiment.

As has already been shown, the Regiment was raised at Mominabad in 1826 by Captain Sir John Bury Gordon of Park, Baronet, an officer of the 13th Light Dragoons, which regiment was at that time serving in the Madras Presidency. Sir John Gordon became the first commandant.

Immediately after it was raised, it was employed on field service against the Naiks, Pindarees, Rohillas, Bhils, and other marauders. In fact, during the thirty odd years which intervened till the Great Mutiny, it, in common with the other Contingent Cavalry regiments, was seldom left any leisure for peaceful pursuits.

The record of most of these small expeditions is lost; no doubt at the time they were too common to be deemed worthy of notice, but a certain number are traceable, and are interesting as showing what was accomplished, and the training by which the Hyderabad Contingent attained their brilliant reputation as Irregular Cavalry.

The first station of the 4th Nizam's Cavalry appears

to have been Bolarum. In June, 1826, a troop of the regiment was sent towards Tikalpali in pursuit of the rebel Narsingreddi; and later in the same year, a squadron was employed to restore order in the districts south-west of Bolarum.

Shortly after a rebel with his followers took possession of the fortress of Dhabee, and began to harry the surrounding country. The 4th Cavalry, under Captain John Sutherland, notwithstanding its recent formation, besieged the fort with great vigour, the sowars fighting on foot. After repeated assaults, the fort and whole garrison were captured.

On 12th October, 1827, a battalion of infantry, one squadron of the 4th Cavalry, and two six-pounder guns, marched from Bolarum and reduced to obedience the Killahdar of Moodgul, on the south-west frontier of the Nizam's territory, who had broken out in rebellion. A little later other detachments of the regiment broke up a body of marauders at Beebeepet; a bandit leader, named Kandareddi, and his gang at Alur; and three Naik Zemindars, who had established themselves with their followers in their forts, and were committing depredations in the surrounding country. The whole regiment also about this time distinguished itself in a series of stubborn encounters with the Bhils, who were accustomed to raid from Ajunta and the neighbouring hills.

After a review of the troops of the Hyderabad Division, on the morning of 15th December, 1828, the Resident issued the following complimentary order:—

"The Resident observed with great satisfaction the efficient state of the 4th Regiment of Cavalry, which, notwithstanding the unfavourable circumstances under which it was originally formed, has attained a degree of discipline which qualified it to take a distinguished part in the manœuvres of the regular troops, and which could only have been the result of a careful and judicious system of instruction by the officer commanding it."

On 13th November, 1829, a serious disturbance against the Nizam's authority took place in Hyderabad City, headed by the Nizam's youngest brother. Sir John Gordon, taking his own regiment, and some infantry and artillery of the Hyderabad Division, made a rapid march from Bolarum to Hyderabad City, and promptly brought the prince to terms. The Resident, in a General Order, warmly acknowledged the prompt and efficient way Sir John Gordon, and the officers and men under his command, had performed this duty, and thereby enhanced that reputation for discipline, skill, and gallantry which their former achievements had acquired.

During 1831, a troop was employed from Aurungabad in protecting the frontier against a party of Ramosis, who had escaped from British territory about Nagpur.

One of the rules of the Hyderabad Contingent Cavalry was, that no men wounded in the back need ever expect promotion. An interesting case occurred during 1831, showing how failures were dealt with in the regiment. During April and May, a detachment was sent towards Wargaon, Dounat, and Seolee; they, however, failed to accomplish the object of their expedition. A *panchayat*, consisting of the risaldar and jemadars, at once assembled, held a searching investigation, and decided that, with the exception of nine horsemen, officers and men had failed in the performance of their duty. The Resident concurred in the finding, and the whole number were paid up and dismissed on the spot.

In 1834, one troop from Mominabad was employed, from 3rd May to 29th November, in suppressing a band of freebooters, who were plundering villages in the Latoor *pergunnah*.

On 4th March, 1835, two half troops marched from Hingoli, one in the direction of Rissood, the other towards Sirpur, in pursuit of Balapoliah, a noted bandit, who was committing depredations in the vicinity of those places.

A troop, under command of Captain Malcolm, was despatched from Hingoli, on 8th February, 1836, to suppress freebooters. On the 10th *idem* they captured, at Datara, Kunda Rao and nine other adherents of the notorious Balpoliah. Another troop, with two companies of infantry, was sent to escort 250 Rohillas from Maiker out of the Nizam's country.

On 5th October, 1836, the headquarters of the 4th Cavalry, under Captain Byam, with Ressaidars Mir Mahbub Ali and Mirza Wazir Ali Beg, marched from Bolarum to co-operate with the Company's troops against Raja Bhirbhul Pathoor, who had rebelled and placed himself at the head of a body of Varriahs and Ghonds in Gumsur.

It was of vital importance that they should arrive as early as possible, as the force was deficient in cavalry; Captain Byam and his men reached their destination on 5th November, with their horses fresh and fit for action, having marched 588 miles in 31 days. This march was all the more remarkable, as it was made across the hills and jungles of Gumsur and Burhanpur, which were so dense as to be penetrated with difficulty by infantry.

The rebellious Raja was completely defeated in several engagements.

The march and subsequent actions attracted much attention at the time, and was the subject of several complimentary orders. The Commissioner of Gumsur brought to notice "the valuable and effective services of Captain Byam and his men, who have ever been forward when anything was to be done, and have on all occasions merited my warmest acknowledgments."

Brigadier-General Taylor, commanding the Northern Division, also testified to this service in the highest manner, especially remarking on the readiness, activity, and good feeling with which Captain Byam and every man under his command acted on all occasions, however harassing and trying the duty might be, and saying he should take an early opportunity of making known the estimation, in which the Irregular Horse had been held by every officer with whom they had served.

Finally, the Right Honourable the Governor in Council issued a General Order, dated Fort St. George, 4th March, 1837, stating:—"The party of His Highness the Nizam's Horse, under Captain Byam, likewise merits special notice. In order that he might be in time to join before the commencement of hostilities, Captain Byam made a march of 588

Facing Page 40.

CAMEL SOWAR, 1902.

miles in 31 days, and brought his men and horses to the frontier of Gumsur fresh and perfectly efficient; his services and theirs were, during the time they were employed, fatiguing and incessant, but were performed with unwearied zeal and alacrity, greatly to their own credit and to the benefit of the public interests."

This was followed by a letter from the Governor-General in Council expressing his satisfaction, etc., and requesting the Resident "to communicate the approbation of his Lordship in Council."

The 3rd Cavalry, under Captain Davidson, and the right wing of the 4th Cavalry, under Captain Garstin, the whole under Colonel James Blair, left Mominabad about the middle of September, 1839. It was the rainy season, all the rivers were in flood, and marching over the black cotton soil was very difficult. However, they reached Alampur on the 27th September. Here they were employed to co-operate, with Madras troops under General Wilson, against the Raja of Kurnool, who had revolted. Colonel Blair and the two regiments held the fords of the Tungabudra and Kistna Rivers till 12th November, when the fort of Kurnool surrendered. The regiments then returned to Mominabad, leaving one squadron to temporarily garrison Kurnool, and escort in irons to Hyderabad 338 Rohillas they had captured. The services of the 3rd and 4th Cavalry on this occasion, were recognised by the Resident in a General Order, in which all ranks were thanked for the promptitude with which they started, and marched through a country rendered extremely difficult by the inclemency of the weather. Their conduct was described as being "such as to uphold the high reputation for efficiency and readiness for active service which has ever characterized the Nizam's Cavalry."

On 5th September, 1841, the regiment marched from Mominabad with the right wing of the 1st Cavalry, the whole under Brigadier Blair. Afzalpur was reached on the 12th, and the fort of Barurgi, which had been seized by a body of Arabs and others, was invested. The garrison surrendered on the 23rd.

Immediately on the conclusion of this expedition, a field force, under Brigadier Blair, from the Hyderabad Division, consisting of three troops of the 3rd Cavalry, the 6th Infantry, and six guns, reinforced by the 1st Cavalry and 4th Infantry, moved out against Arab mercenaries, under one Kaheran, who had broken out into open mutiny, marched through the south-west portion of the Nizam's Dominions, crossed the border, and taken and plundered the fort of Badami. The 4th Cavalry marched in five days from Hingoli to Bolarum* to co-operate. There was a fight at Amagundi, and Badami was retaken. The field force was then broken up into detachments, which were in constant pursuit of parties of insurgent Arabs, Rohillas, and other mercenary troops, until the end of December. The services of the regiment on this occasion were favourably noticed by the General.

In 1843, under instructions from the Government of India, the following troops, under the command

* A distance of just on 200 miles.

of Brigadier Bagnold, marched for service towards Balapur: 1st Cavalry, 200 sabres, from Ellichpur and Hingoli, 3rd Cavalry, 180 sabres, from Aurungabad, 4th Cavalry, 220 sabres, from Mominabad, 3rd and 8th Infantry, and two guns from each of the 2nd and 4th companies of artillery. This force reduced the whole country to order.

Shortly after, the 4th Cavalry, under Captain Strange, marched from Aurungabad against predatory Rohillas, who had assembled in force at Wurronee in the Deoran *pergunnah*. Marching 100 miles in three days, they surprised the Rohillas, captured 300, and drove them beyond the Burhanpur Frontier.

The Headquarters of the Cavalry Division, with the 1st Cavalry, under Brigadier Beatson, and the Aurungabad Division, consisting of the 4th Cavalry, 2nd Company of Artillery, and 5th and 8th Infantry, were ordered to take the field, on 4th May, 1849, for the purpose of clearing the country of roving bands of Rohillas and dacoits; and also against Mokut Rao, who had made his appearance in the Nizam's territory with a band of followers. The force captured 97 Rohillas and 34 dacoits at Arin, on 21st May, and 59 Rohillas on 28th May.

The Fort of Rai Mhow having been seized by a party of outlaws under Sidi Nasirullah, Brigadier Beatson was directed to proceed against this individual, and take measures for the transfer of the fort to the Nizam's officials. The Brigadier accordingly marched with the 2nd Cavalry, and two squadrons of the 4th Cavalry from Mominabad, 23rd November, 1850. Reinforcements of the 3rd Company of Artillery, and five companies of the 4th Infantry, from Hingoli, were received. The force arrived before Rai Mhow 28th November, and immediately invested the fort; but on the 5th December, while awaiting siege guns, which were being despatched from Aurungabad, and a wing of the 6th Infantry from Bolarum, the enemy surrendered.

At this same time a similar detachment from Aurungabad, consisting of three companies 8th Infantry, a squadron 4th Cavalry, and two guns, 2nd Company of Artillery, under the command of Captain Peyton, was employed against Rohillas, and in quelling disturbances in the Mulkapur district.

Dharoor, one of the strongest forts in the Nizam's dominions, having been seized by a party of Afghans, Brigadier Beatson was directed to recapture the fort with the least possible delay; and to adopt such measures as would frustrate any attempt at escape by the enemy, whose accumulated crimes made it necessary that they should be made an example of. The Brigadier marched against Dharoor with four guns of the 3rd Company of Artillery, five siege guns of the 2nd Company of Artillery, the 2nd Cavalry, two squadrons 4th Cavalry, and a squadron 5th Cavalry.

The force was joined subsequently by the 3rd Cavalry. Batteries having been erected, fire was opened 27th January, 1851, and a practicable breach effected on 4th February, but just as the dismounted cavalry were preparing to carry the place by storm, the enemy surrendered and laid down their arms. Two 18-pounder guns and two companies of the 3rd Infantry from Bolarum, and one 18-pounder gun from Hingoli, were *en route* to Dharoor at the time of the surrender.

Brigadier Beatson issued a highly complimentary order to all the troops engaged, which finished up with: "I have now to add that when I called for 50 volunteers from each of the 2nd Cavalry, and detachments of the 4th and 5th Cavalry, to serve on foot along with the infantry in storming the breach, which the skill of the Artillery had rendered practicable, the whole of the officers and men of the 3rd Cavalry Regiment present begged to be allowed to serve on foot in the assault."

Brigadier Mayne left Mominabad, 8th March, 1852, with 250 men of the 2nd Cavalry, to operate against freebooters in the Oodghir district. He was reinforced by 100 troopers of the 4th Cavalry, from Gulburgah on the 13th, and a detachment of two companies of the 4th Infantry from Hingoli, between the 19th and the 24th March, 1852. This force was in pursuit of Linguppah and other plunderers in various parts of the country. Having captured seven of their strongholds, which were subsequently destroyed, the field force was broken up on 4th April.

Detachments, consisting of 100 troopers from the 2nd Cavalry, under Captain MacIntyre, 100 from the 3rd Cavalry, under Captain Abbott, and 100 from the 4th Cavalry, under Captain Clagett, marched from Mominabad, Hingoli, and Gulburgah respectively, on 1st December, 1852, against the rebel Nursi Rao, and reduced the fort of Balligaon.

A field detachment of a wing of the 4th Cavalry, two guns of the 1st Company of Artillery, and a wing of the 3rd Infantry, under the command of Brigadier Hampton, marched from Bolarum, 26th May, 1853, for the protection of the Raichur Doab district.

During the months of January and February, 1854, a force consisting of a wing of the 3rd Cavalry, three guns of the 3rd Company of Artillery, and a wing of the 4th Infantry from Hingoli, a squadron of the 4th Cavalry from Bolarum, under Lieutenant Hastings Fraser, and a wing of the 1st Infantry from Wurrungul under Captain Hare, was employed in the central districts of the Nizam's territories, under the command of Captain Hare. The force was present at the affairs of Chillergeer, Sirpoor, Edlabad, and Nubbeepet; 450 Rohillas were taken prisoners.

Brigadier Mayne had the pleasure of communicating to the Resident, in a letter dated 20th May, 1854, that the native officers and men of the 4th Cavalry under his command, " thinking it possible that, in consequence of the war between Russia and Turkey, they may be found useful across the seas, had begged leave to volunteer their services to proceed to Turkey or any part of the world in which Government may wish to employ them."

A flying column, consisting of a squadron of the 4th Cavalry, four companies of the 4th Infantry, and three guns of the 3rd Company of Artillery, under Captain Doria, commandant 4th Cavalry, left Hingoli on the 9th September, 1855, against a body of about 1,000 Rohillas, who were in the vicinity of Degloor. The cavalry covered a distance of 160 miles in four days. On 12th September, 1855, the enemy were attacked and defeated at Banda Koonta, where they had taken up a position among some hills and caves. The gallant conduct of the troops under Captain Doria was brought to the notice of the Governor-General by the Resident. Many prisoners were captured; their arms were sold by auction, and the proceeds divided as prize money among the troops.

A detachment of a troop of the 4th Cavalry, two guns 3rd Company of Artillery, and two companies 4th Infantry, marched from Hingoli, on the 28th September, 1855, under the command of Captain Daniel, and quelled an outbreak at Parbhani.

In May, 1856, a squadron 4th Cavalry, with four companies 4th Infantry, the whole under the command of Captain Doria, 4th Cavalry, were on service towards Mungloor.

The following year the Great Mutiny in India broke out. A special chapter has been devoted to the narration of the loyal and brilliant services of the Regiment under Sir Hugh Rose, in the Central India Campaign.

After the capture of Gwalior the 4th Cavalry returned to Hingoli; but, in November, 1858, within a month of their return, the Hyderabad Contingent were again ordered to take the field to prevent Tantia Topee, Rao Sahib (Nana's nephew), and the Nawab of Banda, from crossing south and creating a disturbance in the Deccan.

The force included the 4th Cavalry, now commanded by Major H. D. Abbott, C.B., Major Murray and Lieutenant Hastings Fraser having been transferred to the 1st Cavalry, and Captain Nightingale to the 3rd.

At the commencement an advance was made towards Amraoti, Berar, but on arrival there, it was found that the rebels had changed their plans, and were moving from the Satpura Hills towards Asirgurh and Nimbhar. The Hyderabad Contingent Force, leaving Amraoti, crossed the Purnah River to Burhanpur, and advanced to Akbarpar, pursuing the rebels so closely that they could halt nowhere. By these prompt and effective measures the whole district of Nimbhar, from the Nerbudda to Burhanpur, as well as the Nizam's Dominions, were kept peaceful, and the rebels and other marauders prevented from plundering.

The Field Force was under the command of Brigadier Hill, and consisted of the 1st, 3rd, and 4th Cavalry, three troops of the 2nd Cavalry, six companies from each of the 3rd, 5th, and 6th Infantry, four guns of the 1st Company of Artillery, and two guns from each of the 2nd and 4th Companies of Artillery.

A field detachment of the Hyderabad Subsidiary Force left Secunderabad, on the 16th November, to co-operate.

In addition to the pursuit of Tantia Topee, the Field Force was also employed against bodies of Rohillas and marauding Arabs. It attacked and defeated, on 15th January, 1859, a body of Rohillas, who had taken possession of the village and fort of Chichambah, and returned to cantonments on 22nd March, after having destroyed a number of fortified places captured from the enemy. The loss in this

DUFFADAR, DRILL ORDER.

action was nine killed (including one British officer), and twenty-three wounded.

During February and March, 1859, a mixed force of Madras and Hyderabad Contingent troops, the latter being detachments of the 2nd and 4th Cavalry, 1st Infantry, and two guns 1st Company of Artillery, all under the command of Lieutenant-Colonel Orr, were employed in pursuit of an Arab named Shaik Ahmed. Subsequently they operated, at Narrain, against the Rohillas, who had plundered Nellingah.

A force, consisting of the 1st troop 4th Cavalry, one hundred and fifty men 3rd Infantry, and the 1st Company of Artillery, under the command of Major Abbott, left Hingoli, 31st October 1859, on service against a party of Rohillas. These were attacked and defeated at Jintoor on 3rd November, and the force returned to Hingoli 7th November, having one trooper killed and two wounded, also six infantrymen wounded.

This action may be said to conclude the services of the 4th Cavalry in the pacification of the Deccan, but long after this the regiments of the Hyderabad Contingent were often called upon to suppress bands of dacoits.

1860.—The Regiment left Hingoli 6th January, and arrived at Aurungabad 17th *idem*. On 1st December, the cantonment of Aurungabad having been made over to the Bombay Presidency, the 4th Cavalry marched out of Aurungabad, with the rest of the garrison, towards Buldana in Berar, in search of a suitable site for a new cantonment. Yelgaum was selected for this purpose.

1861.—The Regiment arrived at Yelgaum on 26th January. In April, Risaldar-Major Muhammed Oomer Khan died, deeply regretted by all. When the Mutiny broke out, in 1857, he was on furlough in Oude, and did good service to Government, and also saved the lives of several British officers and ladies. He was rewarded for this with a valuable *jaghir*. In May, Risaldar-Major Shah Mirza Beg, supernumerary in the 3rd Cavalry, was transferred to the 4th, vice Muhammed Oomer Khan, deceased.

Shah Mirza Beg had received the 2nd Class of the Order of British India for gallantry in the action at Dhar, 22nd October, 1857. He had been distinguished for valour on many previous occasions, and later in the Central India Campaign, and against the Rohillas fully maintained his fine reputation.

On 3rd June, the Regiment returned to Aurungabad, the station having been again made over to the Hyderabad Contingent. This same year Brigadier Hill resigned the command of the Contingent, and was succeeded by Brigadier Lumsden.

1863.—The 4th Cavalry left Aurungabad on relief 16th November, and arrived at Mominabad 28th November.

1865.—Colonel Abbott, C.B., officiated in command of the Hyderabad Contingent from 13th May, 1865, to 9th December, 1866, vice Brigadier Lumsden, on sick leave to Europe. During this time Major Dowker, C.B., officiated in command of the Regiment.

1867.—On 4th September, the 4th Cavalry volunteered for service in Abyssinia.

1868.—The Regiment marched from Mominabad 18th November, and arrived at Bolarum 6th December. Colonel Abbott assumed command of the Hyderabad Contingent, with the rank of brigadier-general, vice Lumsden, vacated on promotion, 28th December.

1869.—On 23rd March, Captain A. Acheson Johnson assumed command of the Regiment, vice Abbott.

1873.—The Regiment left Bolarum 15th November, and arrived at Hingoli 11th December, sending a detachment of a squadron to Ellichpur.

1874.—In December, Brigadier-General T. Wright, C.B., from the 11th Bengal Lancers, assumed command of the Contingent, vice Abbott.

1875.—During January the Regiment took part in the first camp of exercise at Jalna.

1877.—Major C. J. O. Fitzgerald took over command of the Regiment from 18th March, consequent on the departure of Major A. A. Johnson, on furlough to England.

1878.—There was a bad outbreak of cholera in the Regiment, which had to move out into camp from 13th May to 7th June.

1879.—Major A. Johnson returned from furlough on 5th March, and took over command from Major Fitzgerald.

1880.—The 4th Cavalry arrived at Aurungabad on relief 3rd January. A detachment was temporarily sent to Mominabad.

1884.—Lieutenant-Colonel Walker took over command of the Regiment, vice Colonel Johnson, retired.

There was another camp of exercise at Jalna, from 29th December till 23rd January following.

1885.—The Regiment changed stations to Mominabad, arriving there 3rd February. The 4th Cavalry volunteered for service in the Soudan, and was thanked officially by the Secretary to the Government of India. Major Cummins, now second-in-command, proceeded on active service to Suakim. He was accompanied by Duffadar Muhammed Ali Khan; both were present at the Battle of Tamai.

1886.—On 2nd September, the 4th Cavalry arrived at Bolarum to relieve the 3rd Cavalry, ordered on field service to Upper Burma. Two assistant salootris were lent to the 3rd Cavalry. Major Cummins also proceeded on this expedition as a transport officer; Duffadar Kesar Singh accompanied him as an orderly.

1887.—Colonel Walker was on furlough to Europe, and Colonel Bell, of the 2nd Cavalry, officiated as Commandant till the return of Major Cummins from the Burmese War. Risaldar-Major Muzaffar Khan proceeded to England, as one of the representative officers of the Indian Army, to attend the Jubilee celebration of Her Majesty Queen Victoria, by whom he was decorated with the Jubilee Medal, and promoted to the 1st Class Order of British India, with the title of Sirdar Bahadur. He was subsequently made a Companion of the Indian Empire. He had served with the Regiment in the Mutiny, 1857, for which he had received the medal. On 19th September, the Regiment was ordered on field service to Burma.

A special chapter has been devoted to the services of the 3rd and 4th Cavalry in the Burmese War, 1887-8-9.

1889.—The Regiment returned to India from the Burmese War, and was stationed at Mominabad; the last squadron arrived there 29th March.

Colonel Walker left for England in May on furlough, pending retirement, and on 9th October Captain H. M. Mason was transferred from the 1st Cavalry as officiating Commandant, vice Lieutenant-Colonel Cummins appointed Commandant of the 2nd Cavalry.

On October 23rd, the 4th Cavalry marched for Bolarum to relieve the 3rd Cavalry there. On arrival it was inspected by General Luck, the Inspector-General of Cavalry, having to fight its way in through the outposts of the Secunderabad Cavalry Brigade and 3rd Cavalry.

The Regiment was now in brigade with the 7th (Queen's Own) Hussars and 1st Madras Lancers.

1890.—In August, medals for the Burma Campaign were presented by Sir Denis FitzPatrick, K.C.S.I., Resident at Hyderabad.

On 1st November, Risaldar-Major Muzaffar Khan, C.I.E., retired, receiving a double pension after a long and unblemished service of nearly 36 years. Ressaidar Wilayet Ali Khan succeeded him as Risaldar-Major.

1891.—From 2nd to 25th January, the 4th Cavalry was at the Secunderabad Camp of Exercise, at which the 1st and 2nd Cavalry Hyderabad Contingent were also present, making, with the 7th Hussars and 1st Madras Lancers, a strength of five cavalry regiments. The Inspector-General of Cavalry attended the camp, and inspected all the corps, a good deal of brigade and divisional drill being carried out under his personal command.

About September the 7th Hussars left for Mhow, and the 19th Hussars, from England, then commanded by Colonel French (now Sir John French), temporarily took their place till relieved in December by the 21st Hussars.

1892.—Risaldar-Major Wilayet Ali Khan died 26th October, from effects of a railway accident at Tundla, when travelling on remount duty. He had done 35 years of long and faithful service, and was extremely popular with all ranks. He was an Indian gentleman of extremely charming manner, and had done much useful work for the Regiment in purchasing remounts; in fact, he had the reputation of being one of the best judges of Arab horses in India, and certainly the class of Arabs in the Regiment at this time was very high. Ressaidar Shaik Muhammed Sultan was promoted risaldar-major in his place.

1893.—In March the Regiment was inspected by the Inspector-General of Cavalry, General Luck. Lieutenant Stotherd left 18th March on special reconnaissance duty across Persia to Russia. Lance-Duffadar Mehndi Husain accompanied him. Duffadar Amir Muhammed Khan was selected to attend the opening of the Imperial Institute in London; he left for England, and returned 16th September. He was promoted Jemadar by order of Her Majesty Queen Victoria.

In October, General Grant, C.B., the new Inspector-General of Cavalry, inspected the Regiment.

1894.—On 25th November, the 4th Lancers left Bolarum for Bedur Camp of Exercise. They were in brigade with the 21st Hussars and 1st Madras Lancers, and in the opposing Cavalry Brigade were the 1st, 2nd, and 3rd Lancers Hyderabad Contingent. At the conclusion of the camp they marched to Hingoli on relief, arriving there 29th December.

1895.—Brigade-Surgeon Lieutenant-Colonel Sargent exchanged, from the 2nd Lancers Hyderabad Contingent, with Surgeon-Major Kellie, who had been medical officer of the Regiment for many years, including the time when on active service in Burma. Colonel Sargent was a very old Hyderabad Contingent officer; and to the great regret of his many friends in the various corps, he died this year at Hingoli.

The Government mules with the Regiment were requisitioned for service with the Chitral Relief Force on 15th April. They left at ten hours' notice, made marches of thirty miles a day to Akola, whence they trained to Agra, where they were equipped and sent on to Nowshera for the campaign. Duffadar Muhammed Amir Khan, and two sowars, Bahadur Khan and Shekh Myboob, accompanied them.

1896.—Risaldar-Major Shaik Muhammed Sultan

left on pension 1st May, and was promoted Bahadur. Ressaidar Abdul Majid Khan succeeded him as risaldar-major.

1897.—In August of this year there was a bad outbreak of cholera in the regimental lines, and among those who died from it was Captain F. J. Nelson. He was deeply regretted by his brother officers and the whole Regiment.

The Government transport mules were mobilised in September for service on the North-West Frontier. Duffadar Govind Rao, Lance-Duffadar Pal Singh, and Sowar Shekh Myboob accompanied them. They took part in the operations on the Samana, in Tirah, the march down the Bara Valley, reopening of the Khyber, and other expeditions. Two salootris, two ward orderlies, and one sowar orderly from the Regiment, also took part in this campaign. They all returned on 27th April following.

1898.—The Regiment was head of the list of Indian cavalry for Army signalling. It was during this year that mules were first bought for the regimental transport instead of ponies.

1899.—The Regiment was first in the Indian Army for Army signalling, and a complimentary order was issued by the Commander-in-Chief.

The reliefs of all the Hyderabad Contingent Cavalry Regiments, which should have taken place this year, were cancelled on account of famine caused by the failure of the monsoon. Owing to scarcity of grass and fodder, the Regiment moved out in October to a camp at Belura on the Pen Gunga River, leaving 100 horses and about 140 men as a depôt at Hingoli.

Facing Page 56.

Jat Sowar (Marching Order). British Officer.
Risaldar Major (Full Dress). Sikh Duffadar.
1911.

Grass was brought on the regimental ponies from the Marwari Forest Reserve, and there was a sufficiency of water standing in the deeper pools in the river bed.

On 31st December, Kote Duffadar Muhammed Amir Khan, Lance-Duffadar Sahib Singh, and Farrier Ramchunder, with eight *saises*, proceeded to South Africa in charge of thirty horses, purchased with complete saddlery by Government, as remounts for the war.

The first place in the Madras Chart and Compass Competition was won by the 4th Lancers.

At the end of December, Major S. M. Mason, second-in-command, died in London the same day as he landed in England. He had been suffering for some time from abscess of the liver, and had been granted leave on medical certificate. This officer had distinguished himself in the Burmese War, and had done excellent service for the Regiment, of which he was for several years adjutant. He was a universal favourite with all who knew him.

1900.—Lieutenant P. F. Newnham was killed in South Africa at the action of Spion Kop, while serving as a volunteer with Thorneycroft's Horse. He behaved with great gallantry, when already badly wounded, taking a rifle and continuing firing till struck by a second bullet that killed him.

The Regiment moved from Belura on 5th March, and took up a fresh camp at Wussumba, four miles from Hingoli. Here good water was obtained by sinking wells in the dry bed of the nullahs. This camp was occupied till 28th June, when sufficient rain had fallen to permit a reoccupation of the lines at Hingoli.

Ressaidar Abdul Karim Khan succeeded Abdul Majid Khan as risaldar-major on 1st May, when the latter retired on pension.

On 25th June, Kote Duffadar Govind Rao and five men left for Calcutta, to join the Expeditionary Force to China for the Relief of the Legations. Sowar Atma Singh also accompanied them as orderly to Captain Stotherd. These men were absent for over a year in North China, and were employed at first as orderlies to Lieutenant-General Sir Alfred Gaselee, commanding the British Forces, and the Headquarter Staff, and afterwards in organising local transport corps.

The Regiment left Hingoli 17th December, and arrived at Aurungabad on relief 23rd December, 1900. On 7th February, 1902, Colonel Mason gave up the regimental command, after having held the appointment, including officiating time, for between twelve and thirteen years. When he took over command, the Regiment was still disorganised from the effects of the Burmese Campaign, and there were in it nearly two hundred young horses, newly purchased to replace casualties. But the material, both men and horses, was excellent, and he made the most of it; after twelve years of unremitting exertion and carefully-thought-out reorganisation in every detail, he made the 4th Lancers one of the best native cavalry regiments in India. Major Wapshare, then second-in-command of the 2nd Lancers Hyderabad Contingent, succeeded him in command of the Regiment. In view of the great changes contemplated in the Hyderabad Contingent, the appointment was made officiating.

Risaldar-Major Abdul Karim Khan, Duffadar Natha Singh, and Sowar Har Lal, proceeded to Delhi to represent the Regiment at the Coronation Durbar.

1903.—On the 1st April, the reorganisation of the Hyderabad Contingent was brought into effect. The regiment now became the 30th Lancers, Gordon's Horse, and the Sikh squadron of the 3rd Lancers Hyderabad Contingent was transferred to it, bringing it up to a strength of four squadrons. The following officers were transferred from the 3rd:—Major T. D. Leslie as second-in-command, Captain W. Warner, and Lieutenant Muspratt. Lieutenant-Colonel Wapshare was appointed commandant.

At this time, owing to the interest of Colonel Wapshare, a regimental pack of foxhounds was started. Colonel Wapshare became the first master, and for several years, both at Aurungabad, and later at Bangalore, some excellent sport was shown. The thanks of the Regiment were especially due to His Grace the Duke of Beaufort, who most generously used to present eight or nine couples of the Badminton hounds every year.

This year the three senior ressaidars of the Regiment were promoted risaldars, to place them on the same footing as Indian officers of the Bombay and Bengal Cavalry.

1904.—General Sir Richard Stewart, K.C.B., was appointed honorary colonel of the 30th Lancers. He had served in the 4th Cavalry about the time of the

Mutiny, and was badly wounded in an engagement at Shorapur.

Major T. D. Leslie died at Poona in August, deeply regretted.

From 7th to 10th October, the Regiment was inspected by Major-General Douglas Haig, C.V.O., C.B., Inspector-General of Cavalry, a very searching inspection, which lasted between three and four days. The high appreciation shown by the Inspecting Officer was very gratifying to all ranks. As a reward for efficiency, the Inspector-General recommended that the Regiment should be specially mobilised at Bangalore.

This year the Regiment was first in the Indian Cavalry for the Commander-in-Chief's prize.

1905.—The 30th Lancers left Aurungabad 2nd January, and marched, via Secunderabad, to Bangalore, arriving there 4th March. Here they were brigaded with the 6th Dragoon Guards (The Carabiniers).

The Regiment was selected to carry out a special musketry course for year 1905-6. The instruction was carried out almost entirely by field practices, and some very satisfactory results obtained.

Major G. Rawlins, who had been transferred from the 12th Cavalry, as second-in-command, vice Major Leslie, died at Poona in September from enteric, when serving as Assistant Military Secretary to Sir Archibald Hunter, Commanding the Southern Army.

Major-General Mason was appointed honorary colonel of the Regiment, vice Sir Richard Stewart, who had been killed at Cheltenham by a fall from his horse when returning from hunting.

1906.—In May, Lieutenant-Colonel St. G. L. Steele was transferred from the 2nd Lancers, Gardiner's Horse, as Commandant, vice Lieutenant-Colonel Wapshare, appointed to the Adjutant-General's Staff at Simla. Colonel Wapshare's departure was much regretted by everyone, as he was most popular with all ranks.

The Regiment was 3rd and 4th this year in the competition for the Nanpara Cavalry Cup, which was won by the Guides Cavalry.

Risaldar-Major Abdul Karim Khan received the Order of British India (2nd Class) and the title of Bahadur.

Owing to a severe outbreak of plague in the lines, most of the Regiment was moved out into camp in October.

The 14th Hussars joined the brigade in relief of the Carabiniers, who moved to Mhow.

Major-General Nixon, Inspector-General of Cavalry, inspected the Regiment on 22nd and 23rd November.

The regimental polo team, under the captaincy of Major Barnard, were very successful in December, winning the Novice Cup in the Madras Tournament, and the Hyderabad Contingent Polo Cup at Bolarum.

1907.—In January, there were Cavalry Brigade Manœuvres, followed by combined manœuvres, the whole camp lasting about a month. The Cavalry Brigade consisted of the 14th Hussars, the 30th Lancers, and the Mysore Imperial Service Lancers.

At the birthday levee, held at Simla, His Excellency the Viceroy decorated Kote Duffadar Gurmukh Singh with the silver medal of the Order of the Hospital of St. John of Jerusalem. This was in recognition of the heroism, humanity, and fidelity displayed by him under exceptionally trying circumstances, he having nursed three plague-stricken men back to health, at great personal risk of himself contracting the disease, at a time when he was absent on duty several hundreds of miles from his regimental headquarters, and also from medical aid.

The Regiment had a very successful musketry year, being 1st, 2nd, and 3rd for the Nanpara Cup, winning it from the holders, the Guides Cavalry. They were also 2nd and 4th for His Excellency the Commander-in-Chief's competitions, open to all regiments of cavalry and infantry in the Indian Army.

The 30th Lancers left Bangalore 22nd November, and marched by Bellary and Ahmednagar to Burhanpur, arriving there on 5th February, 1908. Thence, owing to scarcity of provisions on the road, they were railed to Jhansi, arriving there 6th and 7th February.

1908.—The Nanpara Cup was again won. Major-General Grover, C.B., Inspector-General of Cavalry, inspected the Regiment on 24th and 25th November.

1909.—On Proclamation parade, 1st January, Lance-Naik Sher Singh and Sowars Abdul Shahur Khan, Hoshiar Ali Khan, and Jiwan Singh, were presented with bronze medals and certificates from the Royal Humane Society for a gallant attempt to save Sowar Akbar Ali Khan from drowning in a well at Camp Sanda, on 3rd February, when on the march from Bangalore to Jhansi.

The Nanpara Cup was won for the third time in succession.

Colonel Steele was appointed Assistant Adjutant-General Peshawar Division, and Lieutenant-Colonel W. H. Fasken, from the 10th Lancers, Hodson's Horse, was appointed Commandant of the Regiment, 30th August.

At the end of December, the Regiment marched from Jhansi for Muttra, viâ Agra, to join the large cavalry concentration there.

1910.—During January the 30th Lancers were in brigade with 3rd Skinner's Horse and 39th Central India Horse, and took part in large cavalry manœuvres from Muttra to Gurgaon and the neighbourhood of Delhi. There were thirteen cavalry regiments and four batteries taking part in this concentration, which lasted till the end of January. At the conclusion, the Regiment marched back to Jhansi, arriving there towards the end of February.

This year the Nanpara Cup was won for the fourth time in succession, and the Meerut Cup at the B.P.R.A meeting, against 36 of the best infantry teams in India.

1911.—Risaldar-Major Abdul Karim Khan was selected to proceed to England, as one of the Indian orderly officers for this year to His Majesty King George V. He was accompanied by Duffadar Myboob Khan as his orderly. Major Stotherd and Jemadar Rur Singh were appointed to the Indian

Coronation Contingent; two sowars also accompanied this contingent as orderlies. Major Harbord acted as A.D.C. to Sir Edmund Barrow in London at the same time. Thus the Regiment was represented at the Coronation by two British officers, two Indian officers, one duffadar, and two sowars.

In November, the 30th Lancers marched from Jhansi for Delhi, and were present at His Majesty's Coronation Durbar there, proceeding thence to Ambala on relief. Jemadar Natha Singh was best Indian officer at arms at Delhi, winning the King's gold medal. Trumpeter Wariam Singh was one of the Herald's trumpeters at the Durbar. The silver trumpet, banner, and coat he used were afterwards kept by the Regiment.

On arrival at Ambala the Regiment was brigaded with the King's Dragoon Guards, and 9th Hodson's Horse.

CHAPTER III.

THE INDIAN MUTINY—OPERATIONS OF THE MALWA FIELD FORCE—THE CAMPAIGN IN CENTRAL INDIA —THE GWALIOR CAMPAIGN.

IN May, 1857, the great Indian Mutiny broke out, and the flames of rebellion spread rapidly towards the South; the whole of Central India became involved, and was lost for the time being to the British Government. Then it was that the fidelity of the Hyderabad Contingent prevented the revolt spreading to the Nizam's Dominions, and carrying with it the whole of the South of India. The Hyderabad Contingent were the only local corps in India at that time which remained loyal, and not only remained loyal, but did brilliant service in the field in aiding to suppress the Mutiny in Central India, and restore the British supremacy there.

On 8th June, 1857, a force consisting of—

Five troops 14th Light Dragoons,
One battery Bombay European Horse Artillery,
25th Bombay Native Infantry,
A pontoon train,

left Poona and marched, via Ahmednagar, to Aurungabad, where it quelled some incipient disaffection in the 1st Cavalry. It then crossed the

Mutiny, and was badly wounded in an engagement at Shorapur.

Major T. D. Leslie died at Poona in August, deeply regretted.

From 7th to 10th October, the Regiment was inspected by Major-General Douglas Haig, C.V.O., C.B., Inspector-General of Cavalry, a very searching inspection, which lasted between three and four days. The high appreciation shown by the Inspecting Officer was very gratifying to all ranks. As a reward for efficiency, the Inspector-General recommended that the Regiment should be specially mobilised at Bangalore.

This year the Regiment was first in the Indian Cavalry for the Commander-in-Chief's prize.

1905.—The 30th Lancers left Aurungabad 2nd January, and marched, via Secunderabad, to Bangalore, arriving there 4th March. Here they were brigaded with the 6th Dragoon Guards (The Carabiniers).

The Regiment was selected to carry out a special musketry course for year 1905-6. The instruction was carried out almost entirely by field practices, and some very satisfactory results obtained.

Major G. Rawlins, who had been transferred from the 12th Cavalry, as second-in-command, vice Major Leslie, died at Poona in September from enteric, when serving as Assistant Military Secretary to Sir Archibald Hunter, Commanding the Southern Army.

Major-General Mason was appointed honorary colonel of the Regiment, vice Sir Richard Stewart, brass guns, posted on a hill south of the fort; but the guns were charged and captured and turned on their late owners, who were quickly driven into the fort, leaving numbers dead on the field. The 3rd Cavalry made a brilliant charge, losing one jemadar and one trooper killed, the risaldar-major severely wounded, and four troopers slightly wounded. Major Gall, of the 14th Light Dragoons, drew attention to the gallantry of the men of the 3rd Cavalry, who, he said, "all proved themselves true and gallant soldiers, men, indeed, with whom I am proud to serve under the same flag." The fort was now invested and a breaching battery established. The breach was practicable by the night of the 31st, but the enemy, not awaiting the attack of the storming party, evacuated the place and retired under cover of darkness.

After the conclusion of the siege of Dhar, the Malwa column was joined by the Hyderabad Contingent Field Force, under the command of Major W. A. Orr, Commandant of the 1st Company of Artillery, Hyderabad Contingent.

This Field Force had assembled at Edlabad on 22nd July, 1857, and consisted of twelve field guns, 400 Cavalry, and 1,200 Infantry, made up as follows:—

From Ellichpur—A wing of the 5th Infantry, under Captain G. Hare, four six-pounder guns of the 2nd Company of Artillery, and the left wing of 4th Cavalry, under Lieutenant Hastings Fraser.

From Hingoli—The Headquarters and right wing of the 4th Cavalry, under Captain Murray, four six-pounder guns, 1st Company of Artillery, and a detachment of the 3rd Infantry.

From Aurungabad—Six companies of the 3rd Infantry, under Captain Sinclair, four guns of the 4th Company of Artillery, the Headquarters of the 1st Cavalry, under Captain H. D. Abbott, and a detachment of the 4th Cavalry.

Lieutenant F. Samwell, officiating Adjutant of the 4th Cavalry, was appointed Staff Officer to the Force.

Captain Murray commanded the 4th Cavalry throughout the ensuing campaign, the Commandant, G. Nightingale, being on furlough to England.

The advance was delayed at Edlabad for some time, owing to heavy rains and a severe outbreak of cholera among the troops in camp. At length, in October, Major Orr moved forward to Hoshangabad, where he placed himself in communication with the Governor-General's Agent in Central India, and the Commissioner of Nagpur, and first undertook to suppress the insurrection in the Hoshangabad district, where some rebel Thakurs had broken out and plundered several villages. With this object he marched against Piplia Fort, which he was informed was held by Dowlut Singh, the rebel Thakoor, of Raghooghur, who had plundered Satwas and Nimawar. This man was the acknowledged leader of the insurgent bands in the district, and a reward was offered by Government for his capture. He was said to be accompanied by five hundred of his followers and a Thakur, named Suchait Singh, who had long been assisting him with men and supplies.

With a view to effecting a surprise, and seizing Dowlut Singh, Major Orr, taking the 4th Cavalry, made a forced march on the place, and surrounded it on 27th October.

The enemy at once opened fire, but the 4th Cavalry held them till the rest of the force came up. The artillery got into position, and shelled the fort for some time, when the enemy attempted to parley, with a view to gaining time and escaping under cover of darkness. To frustrate their plan, the infantry, under Captain Sinclair, were ordered to advance and force the gate. This was successfully accomplished, and all those found inside were killed, resisting to the last. Those who attempted to escape by throwing themselves from the walls, were either shot or made prisoners by the Cavalry.

Thakur Bhowani Singh was taken prisoner with 42 of his band, whilst his brother, Suchait Singh, with another brother, an uncle, and some relatives—all men of consequence—were killed.

Dowlut Singh, unfortunately, escaped, as he had left Piplia during the night of the 26th, with a large body of his followers, apparently with the intention of disputing the advance of Orr's column at a *ghaut*. His party was challenged by the advanced guard of the 4th Cavalry, but in the dark, and owing to thick jungle, they effected their escape. The Cavalry, however, captured thirty-two of his followers.

Major Orr, in his despatch, especially thanked, among other officers, Captain Abbott, 1st Cavalry, Captain Murray, Lieutenants Hastings Fraser and Samwell, and Surgeon Orr, all of the 4th Cavalry.

The force now marched to Mhow, but hearing of the fighting at Dhar, they only halted there six hours, and then made one long forced march and joined Brigadier Stuart's column there, just after the fall of the place.

Having demolished the fort at Dhar, the British force marched through Western Malwa towards Mandesur, in pursuit of the rebels, who, on November 8th, attacked the cantonment of Mehidpur, which was garrisoned by a native levy, under British officers. The majority of the garrison went over to the rebels, while the British officers, with the loss of one of their number, succeeded in escaping to the camp of the Malwa Field Force, accompanied by half a troop of Cavalry, who had remained faithful.

On receipt of this news by the arrival of the fugitives in camp, a force consisting of 337 sabres of the 1st, 3rd, and 4th Cavalry, Hyderabad Contingent, under command of Major Orr, was sent to attack the rebel garrison at Mehidpur. Major Orr found Mehidpur evacuated by the rebels, who had retreated by the road to Kesari, taking with them the whole of the guns of the Mehidpur Contingent, with the exception of one 12-pounder siege gun. Proceeding at once in pursuit, Major Orr found another 12-pounder on the road about two miles from the cantonment, and in the bed of the Sipra River were discovered the two 12-pounder howitzers, complete with waggons, and two native guns of considerable calibre, as well as a number of carts containing stores. Leaving a guard in charge of these, Major Orr proceeded with the column, and, on approaching Rawal, received intelligence that a large party of rebels, amounting to some four or five hundred men, with two guns in position loaded with grape, were near the village. On nearing Rawal he divided his force into two portions, sending one division, under command of Captain Abbott, to attack from the right, whilst the other, under command of Captain Murray, accompanied him and advanced from the left.

The troops formed line as they advanced at the gallop, and, charging the two guns, each division received the fire of one, and cut down the gunners. The rebels broke up, but continued to fight to the last with much obstinacy and determination, losing about 100 killed and 74 prisoners.

The men had been in their saddles since 4 a.m., and this gallant fight did not end till sunset. In his report, Major Orr brought to notice the willing zeal and alacrity of the entire force on this expedition, as well as the gallantry which marked its close, and brought to notice the services of the British officers: Captain Abbott*, 1st Cavalry, Captain Murray, 4th Cavalry, Lieutenant Clerk, 3rd Cavalry, Lieutenant Hastings Fraser, 4th Cavalry, and Lieutenant A. A. Johnson,* 1st Cavalry, as well as Lieutenant Samwell, 4th Cavalry, Staff Officer to the Hyderabad Contingent Field Force, who was dangerously wounded in the abdomen. He also expressed himself "indebted to Surgeon Orr, 4th Cavalry, and to Assistant-Surgeon Sanderson, 1st

* In later years commandants of the 4th Cavalry.

Cavalry, for the active share they took in the action, as well as for the care of and attention to the wounded."

These officers all took a most active part in the fight, themselves sabring many of the enemy, and showing a gallant example to their men; Captain Murray and Lieutenant Clerk each had a horse killed. Major Orr also particularly brought to notice the gallantry shown by Risaldar-Major Mirza Zulfikar Ali Beg and Acting Risaldar Hamil Khan, both 4th Cavalry; Ressaidars Murtaza Khan, 3rd Cavalry; Alanddin Khan, 1st Cavalry; Jemadars Sikander Ali Beg, 3rd Cavalry, and Sahib Singh, 1st Cavalry; as well as a number of men.

It is related, that when the despatch recounting this affair reached Durand, he handed it over to Major Gall to read to the 14th Light Dragoons and 86th Foot, so that any doubt remaining in their minds as to the loyalty of the Hyderabad Contingent might be finally dissipated.

The news of this gallant exploit had the desired effect. Henceforth throughout the arduous months of campaigning which followed, the British troops placed implicit and well merited trust in their Indian comrades in arms.

Lieutenant Samwell was invalided home, and was succeeded as Staff Officer by Lieutenant Hastings Fraser, who held this appointment till the end of the campaign.

Brigadier Stuart arrived at Rawal with the remainder of the column, including the rest of the Hyderabad Contingent, and, continuing his march, reached the Chambal River on November 19th; thence arrived unopposed in the vicinity of Mandesur at 9 a.m. on the morning of the 21st, and encamped about two miles from the city. About three o'clock in the afternoon the rebels, under Firoz Shah, advanced in force with standards flying, threatening the British flanks and centre. On approaching the right front, they were charged most gallantly by a picquet of the 14th Light Dragoons, supported by the 3rd Cavalry under Captain Orr, and driven back with heavy loss. At the same time the attack on the left was repulsed by the Hyderabad Contingent Field Force, who cleared the enemy out of a village they had occupied and retained possession of it themselves. The 4th Cavalry in this action took one of the enemy's standards, marked with a crescent and bloody hand. The troops passed the whole of the ensuing night standing to their arms.*

Neemuch was still held by the British and besieged by the rebels, so next morning, in order to cut the enemy's communications between that place and Mandesur, Brigadier Stuart moved forward in order of battle, crossing the Bahri ford of the River Saona, about 1,400 yards to the south-west of Mandesur, where he encamped, facing the west of the town, with his flanks well protected by the two branches of the river.

At daybreak, Major Orr had received information

* It afterwards appeared, that the reason why the rebels advanced so boldly from behind the walls of Mandesur was, that they had been given to understand that the Hyderabad Contingent would go over to them *en masse*. In this they were greatly mistaken.

from the picquets on the Neemuch road, that a party of 200 rebel sowars had left Mandesur to join the enemy at Neemuch, and were just leaving the village of Goraria. These were under the command of Risaldar Hira Singh, of the mutinied Mehidpur Contingent. They were at once pursued by a squadron of the 14th Light Dragoons, under Major Gall, two troops, 1st Cavalry, under Captain Abbott, and two troops, 4th Cavalry, under Captain Murray. After a gallop of five or six miles the hostile force was overtaken about two miles south of Piplia, and most of them cut down. Finding the village strongly held by the enemy's infantry, who displayed many standards the Cavalry returned to camp. Concluding that the infantry seen at Piplia was the advanced guard of the rebel force from Neemuch, Brigadier Stuart moved forward at 8 a.m. on the 23rd, crossed the northern branch of the river, and found the enemy in great force to the north, occupying a very strong position, with six guns on rising ground, with a large mud hut protecting their gunners. Covered by a cloud of skirmishers the British line advanced, the hostile infantry, with green banners flying, moving forward through the intervening millet fields to meet them, whilst their guns at the same time opened fire.

The Brigadier halted his line, and opened fire with his guns at 900 yards, subsequently moving forward and eventually capturing the enemy's guns by a brilliant charge. The rebels then fled towards Goraria, whereupon the 3rd Cavalry swooped down on them and killed numbers of them. A farther advance having been made, Goraria was shelled, and strong infantry picquets were posted all round it.

While this action was in progress, the rear guard was attacked by a strong body of the enemy, who had issued from Mandesur. Major Orr was ordered to send two guns to its assistance; he did so, and also moved down with the whole of his mounted troops, consisting of the 1st and 4th Cavalry; a squadron of the 14th Light Dragoons was already there. This operation having been rapidly effected, Major Orr found the enemy had advanced to a position about 800 yards from the rear guard. He accordingly directed the line to advance; the enemy were driven back by the fire of the guns, and the cavalry then charged the masses of retiring infantry, killing and wounding several hundreds of them, and pursuing them till they took refuge in some gravel pits, whence they again opened fire; here Lieutenant Redmayne, of the 14th, was killed. The infantry now coming up, and the cavalry threatening their rear, the rebels retreated to Mandesur.

About two hours after midnight the enemy made an attempt to break out of Goraria by stealing away between two picquets; they were, however, discovered and driven back into the village.

Next morning, the 24th November, the whole force proceeded to storm Goraria, which was shelled for three hours, and then taken by assault by the infantry. The enemy set fire to the houses, and offered a desperate resistance until evening. Some few surrendered; those who attempted to escape were cut up by the cavalry, and the remainder were slain in the village, this holocaust closing the campaign in Malwa, where the rebellion had been effectually suppressed by the operations just narrated. The rebels evacuated Mandesur, and dispersed in all directions.

Regarding these actions, Brigadier Stuart wrote in his despatches:—

"By the successful operations of the Malwa Field Force and Field Force Hyderabad Contingent, in the vicinity of Mandesur, the Neemuch garrison has been relieved from the assault with which it was threatened; the insurgent rebels have been dispersed from their stronghold in which, for months past, they have been daily collecting all those disaffected of our rule; and peace and order will now, it is to be hoped, be re-established in these districts. . . . . Major Orr, Commanding Field Force, Hyderabad Contingent, most ably co-operated with me on all occasions, and to him and all under his command I am very much indebted. . . . . Of Captain Orr, 3rd Cavalry, Hyderabad Contingent, and the officers and men under his command, I have already had to send the most favourable report to the Resident at Hyderabad. On this last occasion nothing could surpass the bravery shown by all ranks of this Regiment. Captain Orr himself is a first-rate cavalry officer; his daring courage is admired by all, and in every affair in which he is engaged his personal combats are most prominent features."

Colonel Durand added his testimony regarding the splendid services of Major Orr and the officers and men of the Hyderabad Contingent, and Major Orr brought to notice the services of his British officers, including Captains Abbott and Murray, Lieutenants Hastings Fraser and Johnson, and Surgeon J. H. Orr.

The main body of the Malwa Field Force now marched to Indore, which was reached about 14th December. The Hyderabad Contingent Force, under Major Orr, was left for the time at Mandesur, where the wounded and prisoners were also collected; they were kept busy for about a month, apprehending rebels and reducing the district to order. A party of a hundred and twenty of Sindhia's rebel sowars, who had fought at Goraria, were captured, tried, and all shot, except a few who were found not guilty.

In the meantime, the Malwa Field Force, after arrival at Indore, had assisted at the disarming of Holkar's Irregular Cavalry. One thousand six hundred men of Holkar's infantry were also disarmed. On 15th December Sir Robert Hamilton, Agent to the Governor-General, arrived and relieved Lieutenant-Colonel Durand.

On 17th December Major-General Sir Hugh Rose, K.C.B., took over command of the forces in Central India. He had entered the Army in 1820, fought in Syria twenty years later, where he was severely wounded, but captured with his own hand the Pasha of Egypt. Prior to the Crimea War he was appointed to Constantinople, and it was he who took the decisive step of ordering the British fleet to

Turkish waters. In the subsequent campaign he bore a distinguished part.

The plan of campaign provided for the advance of three columns; one from Mhow, under Sir Hugh Rose, to sweep the country up to Kalpi on the Jumna, relieving Saugor and recapturing Jhansi. At the same time a Madras force, under General Whitlock, was to cross Bundelkhund from Jubblepur to Banda, while a third column, under Major-General Roberts, operated in Rajputana.

It was to the first of these columns that the 4th Cavalry belonged. This was now designated the Central India Field Force; the Malwa Field Force was the nucleus of it, and two brigades were formed, the 1st, at Mhow, the 2nd, at Sehore.

The 1st Brigade was commanded by Brigadier C. S. Stuart, of the late Malwa Field Force, and consisted of—

Left wing, 14th Light Dragoons, under Major Gall.

One troop, 3rd Bombay Light Cavalry.

*Two companies 86th Regiment (Royal County Down).

25th Bombay Infantry.

Two batteries European Artillery.

Sappers.

The 2nd Brigade at Sehore was under Brigadier Charles Stewart, C.B., 14th Light Dragoons, and was composed of—

Right wing and Headquarters, 14th Light Dragoons, under Major Scudamore.

* The remainder of the 86th Regiment joined later, on 16th March, the day before the capture of Chanderi.

Headquarters, 3rd Bombay Light Cavalry.

3rd Bombay European Regiment.

24th Bombay Infantry.

One battery, Horse Artillery (Eagle Troop).

One field battery.

Madras and Bombay Sappers.

A siege train, with guns worked in action by drafts from field batteries.

A detachment of all arms of the Hyderabad Contingent, under Captain Hare.

The plan of campaign was, that the brigades should move on parallel lines: the 1st to Chanderi, which was to be captured; the 2nd on Rahatgarh and relieve Saugor. The two brigades were then to form a junction between Chanderi and Jhansi, for the attack on the latter place.

Major Orr, with whom was the greater part of the Hyderabad Contingent Cavalry, at first formed an advanced guard to the 1st Brigade, but it appears afterwards that the duty of keeping up communication between the two brigades also devolved on his cavalry, as they crossed over and assisted the 2nd Brigade in the forcing of the passes. On the whole, their rôle during this phase of the campaign was most often that of independent cavalry.

Major Orr left Mandesur about the end of December, and moved first to Augur, reducing the country to order. Then he preceded the 1st Brigade to Goona, restoring the telegraphic communication, which had been destroyed by the rebels.

The operations of the 2nd Brigade, with which was Sir Hugh Rose himself, will be narrated first. They were delayed at Sehore, awaiting the arrival of the siege train, which at last arrived 15th January. The brigade left the following day, marching first on Rahatgarh, a very strong fortress 25 miles from Saugor, standing upon the extremity of a long, high hill, surrounded by dense jungle. The rocky sides of this eminence are scarped and precipitous, falling at one point sheer to the deep and rapid waters of the Bina River, and allowing of access only by a steep and narrow road. The north face of the fort was covered by a strong wall and a ditch twenty feet wide. On the west it overlooks the town and the road to Saugor, having bastions at intervals along the walls, in the angles, and flanking the gateways.

Before this formidable stronghold Sir Hugh Rose arrived with his 2nd Brigade on 24th January, 1858. The rebels were quickly driven in to the shelter of the fort and town. On the 26th the guns opened fire, and the breaching batteries were completed by the morning of the 27th. By the night of the 28th the breach was found to be practicable, when a body of some 2,000 insurgents, under the Raja of Banpur, made an attempt to raise the siege. These were, however, routed with great confusion, by a detachment of cavalry and horse artillery and the 5th Infantry Hyderabad Contingent.

This so disheartened the garrison that they evacuated the fort in the night, escaping by a precipitous path. They were pursued and a number cut up and taken prisoners; the Hyderabad Contingent Cavalry killed forty. On 30th January the Raja of Banpur was again defeated after some sharp fighting; marching on Saugor, Sir Hugh Rose

Facing Page 80.

SIR HUGH ROSE, K.C.B.
(BARON STRATHNAIRN).

entered that place unopposed on the morning of the 3rd February, and raised the siege, which had lasted for nearly eight months. The garrison consisted of 173 Europeans, including women and children.

Here communication was established with Major Orr, who, as explained above, with a force of the Hyderabad Contingent, mostly cavalry, was operating in advance of the 1st Brigade on the Grand Trunk Road from Mhow to Agra. From Saugor, a mixed detachment of the Hyderabad Contingent, under Captain Hare, captured the fort of Sanoda, 8th February.

Sir Hugh Rose, who had been joined by some reinforcements, advanced on the strong fort of Garhakota, which was invested on 11th February, and stormed the following day, the rebels flying in the greatest confusion, abandoning cattle, stores, and everything. They were pursued by Captain Hare, with half a troop of horse artillery, a troop of 14th Dragoons, and a troop of Hyderabad Contingent Cavalry, who cut up about a hundred of them, mostly rebel sepoys.

The British force then returned to Saugor and prepared for a farther advance; fresh supplies of stores, ammunition, etc., were obtained with difficulty, and it was not until 27th February that they were able to move. This unavoidable delay in the operations inspired the rebels with fresh courage, and they again occupied strong positions in the districts—the forts of Serai and Maraora, and the difficult passes of Narut, Madanpur, and Dhamani in the mountainous ridges which separate the Shahgarh and Saugor districts. Sir Hugh Rose's object was to reach Jhansi as quickly as possible, for the attack on which place it was necessary that the 1st Brigade on the west, and the 2nd Brigade on the east of the Betwa River, should be concentrated. The General decided to take the road to Jhansi through the more open country, getting into communication with the 1st Brigade now due at Goona.

He accordingly ordered the 1st Brigade to move from Goona westwards, and take Chanderi, while he himself, with the 2nd Brigade, forced his way northwards, and crossed the Betwa with the final object of making a combined attack on Jhansi with both brigades.

The pass at Narut was by far the most difficult, and the enemy having concluded that the British force must cross it, had increased its natural difficulties by barricading the road with abattis and parapets, made of large boulders of rock, fifteen feet thick. The Raja of Banpur, described as both enterprising and courageous, held this pass with 10,000 men.

The General, however, decided to take the road to Jhansi, through the more open country, and marched from Saugor on 27th February for Rajwas, a central point whence he could operate against any of the passes.

Just previous to this Major Orr and the Hyderabad Contingent Cavalry, including the 4th, leaving Goona, had advanced to Barole, and on February 18th defeated the rebels near there. He joined Sir Hugh Rose near Rajwas on 1st March, having carried out a reconnaissance and collected a great deal of useful information, which decided the General to select the pass of Madanpur for his point of attack. In order to deceive the enemy, and to prevent the Raja of Banpur coming to the assistance of the Raja of Shahgarh, who held the Madanpur Pass, the British General sent a force of the 14th Light Dragoons and other details to make a feint upon Narut; Major Orr, with the Hyderabad Contingent Cavalry, had already reconnoitred up to the Dhamani Pass, and had a skirmish with the rebels holding it.

Sir Hugh Rose himself marched on the 3rd March against Madanpur. Five hundred cavalry, 200 infantry, and four guns, all of the Hyderabad Contingent, together with a company of the 3rd Bombay Europeans formed the advanced guard; some more guns of the Contingent Artillery, 50 cavalry, and 125 infantry of the Hyderabad Contingent, and a troop of the 14th Light Dragoons formed the rear guard. The ground was difficult, with thick jungle on both sides of the steep, rugged road. As the column approached the pass, it came under fire from the jungle and hillsides. Supported by artillery fire, the infantry drove the rebels back to a hill on the left; this was stormed by the 3rd Europeans, and the enemy forced to retreat to a second line of hills. Finally they were driven successively from all the hills commanding the pass; *Captain Abbott cleared the defile with the Hydera-

* Captain Abbott had been temporarily transferred to command the 3rd Cavalry, vice Captain S. G. G. Orr, invalided home on conclusion of the campaign in Malwa. Lieutenant Dowker had then taken over command of the 1st Cavalry.

bad Contingent Cavalry, hunting the insurgents into the village of Madanpur, which was fortified by a thick masonry *bund*, behind which were six guns in position. Sir Hugh Rose then brought up an 8in. howitzer and some nine-pounder guns, the fire of which quickly drove the rebels from their position behind the *bund*, when they attempted to escape into the thick jungle beyond. Many, however, never reached it, for Major Orr and Captain Abbott, with the Hyderabad Contingent Cavalry, dashed forward in pursuit, killing numbers, mostly mutineers, of the 52nd Bengal Infantry, including the infamous Lal Tribedi, the Havildar Major, who was the instigator of the mutiny in that regiment, and whom they had made their Commanding Officer. Sir Hugh Rose marched his force forward several miles into the open country, and camped at Piprai long after sunset, the troops having commenced moving at 3 a.m. The cavalry were so dispersed in pursuit, that many of them were out all night; however, they all came in next morning; their casualties had been very slight, only two troopers badly wounded.

The British force was thus placed in rear of the passes that barred their way to Jhansi.

Meantime the 1st Brigade had advanced on the formidable stronghold of Chanderi, to which the rebels, defeated by Sir Hugh Rose, had fled in large numbers. The place was invested on the 7th March, and carried by assault on the 17th. None of the Hyderabad Contingent took part in the storming of this place, but many of the fugitives from there were intercepted and cut up the following day by the Hyderabad Contingent Cavalry, under Major Orr, near Tal Bahat, a very strong fort, which they had been detached to capture.

Lieutenant Dowker, with 30 sowars of the 1st Cavalry, carried despatches from Sir Hugh Rose to the 1st Brigade, arriving the day before the storming of Chanderi; he made a long, rapid march, losing one man on the way.

On 10th March, the 2nd Brigade reached Banpur and destroyed the palace of the rebel Raja. Tal Bahat, already captured by the Contingent Cavalry pushed on in advance, was reached 14th March; here news was received that the rebels had taken Barwa Sagar Fort and were besieging Orchha. The brigade again advanced, and, 19th March, camped on the left bank of the Betwa. Moving by what is now known as the Lalitpur road, they reached Simra, 14 miles from Jhansi, on 20th March. The same day a force of horse artillery and cavalry, including 476 sabres from the three regiments of the Hyderabad Contingent, were pushed on forward to reconnoitre Jhansi and invest the place. They surprised and cut to pieces 100 Bundelas, who were on their way to join the rebels in Jhansi. Sir Hugh Rose and the remainder of the 2nd Brigade arrived before the city on 21st March, and halted his troops at a distance of a mile and a half from the fort*, while he and his staff went forward and made a careful reconnaissance of the place, lasting till six that evening. Stretching between the open

* About the high ground where the Brigade parade ground now is.

ground, on which he had halted, and the walls of the city, were the charred ruins of the bungalows, in which had dwelt the Europeans massacred the previous June.

Sir Hugh Rose, in his despatches, describes Jhansi somewhat as follows:—"The great strength of the fort of Jhansi, natural as well as artificial, and its extent, entitle it to a place among fortresses. It stands on an elevated rock, rising out of a plain, and commands the city and surrounding country. It is built of excellent and most massive masonry. The fort is difficult to breach, because composed of granite; its walls vary in thickness from sixteen to twenty feet. It has extensive and elaborate outworks of the same solid construction, with front and flanking embrasures for artillery fire, and loop-holes, of which there were in some places five tiers, for musketry. Guns, placed on the high towers of the fort, commanded the country all around. On one tower (see illustration), called the "white turret," recently raised in height, waved the red standard of the Rani. The fortress is surrounded on all sides by the city, the west and part of the south face excepted. The steepness of the rock protects the west; the fortified city wall springs from the centre of its south face, runs south-east, and ends in a high mound or mamelon, which protects by a flanking fire its south face. The mound was fortified by a strong circular bastion for five guns, round part of which was drawn a ditch, twelve feet deep and fifteen feet broad, of solid masonry. The city is surrounded by a fortified massive wall from six to twelve feet thick, and varying in height from eighteen to thirty feet, with numerous flanking bastions, armed as batteries with ordnance, and loop-holes and a banquette for infantry."

The city itself is about four and a half miles in circumference. Outside the walls are trees, gardens and residences, except parts on the south and east fronts. On the east is a large lake with a huge solid masonry *bund*, below which remains to the present day an extensive densely wooded garden. The water palace of the Rani is on the *bund*. To the south lay the ruined cantonments; temples with their gardens—one the Jokan Bagh, close outside the city wall, below the mamelon above mentioned, was the scene of the massacre of the Europeans, the bodies of the victims being thrown into gravel pits there—and two rocky ridges, the eastmost called the Kapu Tekri, both face the south wall and fort.

The garrison holding this stronghold was estimated at about 12,000, including 400 cavalry, with 30 or 40 guns under a first-rate artilleryman. The Rani herself commanded, and had a great personal effect in animating the garrison. Sir Hugh Rose himself described her as, "the best and bravest military leader of the rebels," and, as such, she stands out in brilliant contra-distinction to such cowardly skulkers as Nana Sahib and Tantia Topee.

One result of the General's long reconnaissance on the 21st March was the decision to take the city before assaulting the fort. The wisdom of the course which he had decided upon was shown, when, after the taking of the city, the fort was abandoned by its defenders. He determined to make his assault from two directions, the east and the south. Four batteries were therefore placed on a rocky knoll near the eastern side of the city, south of the lake and opposite the Orchha gate, to take in reverse the mound and the two walls running from it. The 2nd Brigade took up their position here, and were designated the "Right Attack." The four batteries were ready for action on the morning of the 24th, and on the morning of the 25th the bombardment of the city was commenced.

The chief point to be assaulted, however, was the fortified mound forming part of the city wall towards the south, and called by the besiegers the Mamelon. About four hundred yards distant from it was a rocky ridge; from this the assault upon it was made; the first guns placed there were two 5½in. mortars, on 24th March. The 1st Brigade, with siege guns, marched in 25th March, and were placed here, thus becoming the "Left Attack."

The ground thus occupied by the "Right and Left Attacks" did not, however, account for a third of the perimeter of the city and fort; but by 23rd March Sir Hugh Rose had made a complete investiture of the place, by establishing seven flying cavalry camps round the remainder, viz., the north and west faces and most of the east. One large camp, commanded by Major Gall, with a squadron of the 14th Light Dragoons, was on the most distant side of the town; here also were two nine-pounder guns. Another 14th Light Dragoon camp was posted near the water palace and lake, under Captain Thompson

Facing Page 88.

JHANSI FORT FROM THE SOUTH.
(x The White Turret).

of that regiment; and a third of the same regiment was under Major Scudamore, who was placed in command of the whole investing force of cavalry. The fourth camp was formed by the 3rd Bombay Cavalry, and commanded by Captain Forbes. The three remaining camps were those of the three Hyderabad Contingent Cavalry regiments, and were commanded respectively by Captains Abbott and Murray and Lieutenant Clerk. These three held the western side towards Retribution Hill, and were responsible for watching against the approach of an enemy from the direction of Gwalior. The other four camps similarly guarded the approaches from Cawnpore, Kalpi, and the Betwa. All these camps detached towards the beleaguered city outposts and vedettes, which watched and prevented all issue from the city, day and night. On any attempt to force the line, all the camps were called in to support the threatened point. The roads leading to the city were obstructed by trenches and abattis. Prisoners were taken every night, and one party from Kalpi was captured bringing in a convoy of rockets.

The following extract, from the diary of one of the 14th Light Dragoons, gives a good idea of what the work was:—"We were expected to cover a certain portion of the city, to see that none escaped, or to turn out at any moment, and on any emergency. Consequently, we were never out of harness, sleeping in front of our horses, which were always ready saddled and bridled—never having the bits taken out of their mouths, night or day, except a few at a time for feeding purposes, or to give them a drink in comfort, so that it came harder on the horses than it did on us. As for ourselves, I don't think we were able to change our clothes or have a wash for about a fortnight; yet somehow, in spite of this and the dreadful heat, none of us fell sick, and all of us seemed to enjoy the life we led. . . . The system of investment by the "flying camps" of cavalry was most admirably conducted; it was said a cat couldn't pass their lines. Day after day the same routine was rigidly enforced. No quarter was given: those attempting to escape from the city were cut off by our vedettes and sentries, or attacked by our ambush parties posted at night. There was not a night passed but a large number of prisoners were taken by our cavalry piquets, and many of these were summarily disposed of."

On 25th March the siege was vigorously commenced; during that day several towers, including "the white turret," were destroyed by the siege guns, and the powder magazine in the fort was blown up. While the gunners unceasingly bombarded the city from the east, and the Mamelon from the south, the infantry kept up a galling fire against the rebels who lined the walls. On account of the terrific heat at that season, the ordeal to which the European soldiers were subjected was most severe.

In defending the city the rebels showed great courage and determination, and they afterwards acknowledged that they lost between sixty and seventy men a day. Frequent fires broke out, as houses were ignited by the bursting of shells, and the consternation of the beleaguered people rapidly increased.

On account of the great strength of the Mamelon progress there was slow, but on 29th March the parapets of its bastion were levelled, and its guns silenced, though the firing by both sides was kept up unceasingly. It is said that the parapet of the city wall and the ramparts of the fort, all along, often presented the appearance of a sheet of flame. On the 30th a breach was effected at the Mamelon, but it was promptly and with great bravery stockaded; this was mostly destroyed by the besiegers firing red hot shot.

The breach was, however, not yet practicable, and the bombardment was being vigorously prosecuted, when news arrived that Tantia Topee, a relation and agent of Nana Sahib, with 22,000 men, among whom were five or six regiments of the mutinied Gwalior Contingent, and 28 guns, had arrived at Barwa Sagar. On this situation, Colonel Malleson remarks, "the position of Sir Hugh Rose was full of peril. It required, in a special degree, great daring, a resolute will, the power to take responsibility. A single false step, a solitary error in judgment, might have been fatal. But Sir Hugh Rose was equal to the occasion."

He saw that to withdraw the troops investing the fortress, for the purpose of attacking this new enemy, would be a most dangerous step. So he determined to gather together all the men he could, who were not actually on duty in the siege, and meet Tantia Topee with these, while the siege was continued with unabated vigour by the remainder. Under this arrangement only 1,500 men, of whom 500 were Europeans, were available to meet the 22,000 with Tantia Topee. The force selected was furnished by detachments taken from both brigades. The detachments from the 1st Brigade were led by Brigadier C. S. Stuart, while Sir Hugh Rose himself led those of the 2nd Brigade.

Hoping to bring the enemy to an engagement with the river in his rear, the General marched, at 9 p.m., on the 30th, to the village of Basoba, which commanded the fords at Rajpur and Kolwar, leading over the Betwa from Barwa Sagar. No sign of the enemy being found, he decided that they would not cross the river, while he was close to it, and he therefore withdrew, on the 31st, to camp, leaving only outposts to watch the fords, so as to encourage them to advance. The ruse was successful, the enemy swarmed across the upper ford the same day.

The distance from Jhansi to the fords on the Betwa is just over seven miles. Nearly a mile from the city, a long, stony ridge of hills crosses the road at right angles not far from the rear of the camp of the "Right Attack." From this ridge to the river, the country is a mass of undulating ridges and small rocky hills, covered with long grass and scrub jungle, in many places very thick. The surface of the ground is covered with loose stones, and cut up by large nullahs, and is unfavourable for the movement of cavalry. It is very sparsely inhabited.

On the evening of the 31st, the detachments of the 2nd Brigade were drawn up across the road from the Betwa, half a mile from their camp.

It is most probable that the ridge of hills, above mentioned, which crossed the road, was the ground selected, because not only is it the strongest position commanding the enemy's line of advance, but it must have been of vital importance to prevent the enemy occupying it, as it dominated the camp and rear of the "Right Attack," and was eminently suitable as a point, from which to move round and force a passage into the city through the cavalry camps.

Major Orr and some of the Hyderabad Contingent Cavalry had already been sent along the road to the Betwa, to watch the enemy. With Sir Hugh Rose were:—

14th Light Dragoons, 243 sabres.
Hyderabad Contingent Cavalry, 207 sabres.
Some light field guns.
Detachments of the 86th Regiment, 3rd Europeans, 24th and 25th Bombay Infantry, and Hyderabad Contingent Infantry.
Three siege guns.

The General placed one troop of the 14th Light Dragoons under Captain Need, one troop of the 3rd Contingent Cavalry, under Lieutenant Clerk, and four Horse Artillery guns, on his right. In the centre infantry and guns. On the left flank were two troops 14th Light Dragoons and Lightfoot's Battery. In the 2nd line were infantry and guns in the centre, one weak troop of the 14th Dragoons on the right, and Contingent Cavalry on the left. In front of his position he threw out picquets and vedettes of the 14th Dragoons and Contingent Cavalry.

The British force was not in position till long after dark, and the troops slept on their arms, ready for instant action. Shortly after midnight, a sowar of the

Contingent Cavalry galloped in and reported, that the enemy were crossing the lower ford near Kolwar in great numbers. Their obvious intention was to force a way into Jhansi, by way of the Cawnpore road from Bangaon, through the cavalry camps. To meet this movement, it became necessary to detach the 2nd line to that flank (the left). This was under the command of Brigadier Stuart, and consisted of 40 of the 14th Light Dragoons, 107 sabres of the 1st and 3rd Contingent Cavalry, detachments of the 86th Regiment and 25th Bombay Infantry, and some guns. With Sir Hugh Rose the cavalry left were: the remainder of the 14th Light Dragoons, the 4th Cavalry Hyderabad Contingent, and Lieutenant Clerk's troop of the 3rd Cavalry.

At 4 o'clock on the morning of the 1st April the enemy advanced. The cavalry picquets retired steadily, and closed to either flank to unmask the guns and infantry. Covered by a cloud of skirmishers, the enemy advanced to within 800 yards, completely overlapping the small British line, and hoping apparently to envelop its flanks. Their guns immediately unlimbered and opened fire. The infantry of the British were placed behind some rising ground, and were lying down to avoid the heavy fire of the enemy. Their orders were to advance, as soon as the cavalry and artillery attacks from both flanks were well developed. The detachment of the 24th Bombay Infantry were in support, in place of the second line, detached under Brigadier Stuart, leaving only the 3rd Europeans and Hyderabad Contingent Infantry in front.

Sir Hugh Rose describes the opening of the action as follows:—"Before my line was uncovered, the enemy took ground to his right. I conformed, to prevent his outflanking my left, but very cautiously lest he should draw me away too much to the left, and then fall on my right flank. This was probably his intention, for a body of horse was seen to my right. I halted and fronted, the enemy did the same, and instantly opened a heavy artillery, musket, and matchlock fire on my line from the whole of his front, to which my batteries answered steadily.

"The enemy had taken up an excellent position, a little in rear of a rising ground, which made it difficult to bring an effective fire on him."

The General now ordered two guns of the Eagle Troop of Horse Artillery, under Lieutenant Crowe, to move diagonally to their right and enfilade the enemy's left. In this movement a round shot broke the wheel of one of the guns, but the fire of the other was so rapid and accurate, that the enemy's left was shaken. At the same time, Sir Hugh Rose ordered Lieutenant Clerk and his troop of the 3rd Cavalry to charge the rebel battery, that had disabled Lieutenant Crowe's gun. Three times they attempted this, but were driven back by showers of grape and volleys from the Wilayati matchlockmen, losing some men and horses. Lieutenant Clerk was himself wounded and disabled, but Lieutenant Hastings Fraser, of the 4th Cavalry, immediately took over command of his troop, and continued to lead them during the rest of the battle. An advance of the British cavalry from both flanks then decided the day.

From the left, two troops of the 14th Light Dragoons attacked the enemy's right and doubled them up. On the right, the General placed himself at the head of Captain Need's troop of the 14th, and led them against the enemy's left; the 4th Cavalry formed up on both flanks of the Dragoons, and the troop of the 3rd Cavalry, now led by Lieutenant Hastings Fraser, joined in on the right. Opposed were the enemy's best troops, sepoys, and Wilayatis, who, throwing back their flank on two rocky knolls, received the cavalry charge, four or five deep, with a heavy musketry fire. But the cavalry broke through their line, and, bringing up their right shoulders, took the rebel first line in reverse and routed it, driving them back on the Betwa. At the same time, as prearranged, the infantry sprang to their feet, advanced a few yards, then poured in a volley and charged. The enemy broke and fled in complete disorder, pursued by the cavalry and horse artillery, towards their second line, which, under Tantia Topee in person, was some two miles in rear.

While Sir Hugh Rose had been thus engaged, the second line, under Brigadier Stuart, which had moved round the ridge of hills into the plain on the right of the enemy, to prevent them getting round into Jhansi, had attacked, defeated, and pursued them so hotly, that they left gun after gun in the hands of the victors, and numbers of their men dead or dying on the field.

As regards what followed, Malleson says:—"Tantia beheld in dismay the men of his first line rushing helter-skelter towards him, followed by the three arms of the British in hot pursuit; but he had scarcely realised the fact, when another vision on his flank came to add to his anguish." This vision was the rout of a large body of the enemy, caused by the very successful flank movement executed by Brigadier Stuart. "It had the effect of forcing upon him a prompt decision; the day, he saw, was lost, but there was yet time to save the second line and his remaining guns. The jungle was dry and easily kindled, Tantia Topee at once set fire to it, and, under cover of the smoke and flames, commenced a retreat across the Betwa, hoping to place that river between himself and the pursuers. His infantry and horsemen led the retreat, his guns covered it."

But it was all in vain, the 14th Light Dragoons, Hyderabad Contingent Cavalry, and Eagle Troop Horse Artillery, dashed at a gallop through the burning jungle, cutting up hundreds of the rebels in their onward course. The flying enemy often rallied, and many hand-to-hand fights took place; but the farther the pursuit continued the thinner and fewer these rallying masses became. The pursuit did not cease till two troops of the 14th, and the Hyderabad Contingent Cavalry had actually crossed the Betwa; here they became exposed to the heavy fire of the enemy, both in crossing the ford and also in ascending the steep road on the opposite bank. Eighteen guns, two standards, and large stores of ammunition were captured, and also some elephants. The pursuit ended with the capture of the last gun, an 18-pounder, about a mile and a half on the other side of the Betwa. Horses and men were completely exhausted by forty-eight hours' incessant marching and fighting, and they had a nine-mile march back to camp. Malleson says 1,500 rebels were killed and wounded that day. The total British losses were 81 men and 29 horses, of which 15 men and 13 horses were killed. The 4th Cavalry lost two men killed, and among the wounded was Jemadar Sayyid Nur Ali.

The remnant of the enemy, the Peishwa's army as they called themselves, with Tantia Topee, escaped to Kalpi.

Lieutenant Hastings Fraser and Ressaidar Sikander Ali Beg, 4th Cavalry, were both mentioned in despatches for gallantry in this action. The former engaged and killed three of the enemy with his own hand.

Sir Hugh Rose determined to take advantage of the discouragement, which it was well-known the defeat of the relieving army would undoubtedly produce on the minds of the Rani and her garrison, and the siege was prosecuted with renewed vigour. On the 2nd of April the breach was reported practicable, and all the orders for storming were issued. A 24-pound howitzer had been placed in front of the Jokun Bagh, just under the Mamelon (close alongside the gravel pits where lay the bodies of the massacred Europeans), to enfilade and clear, during the night, the wall from the mound to the fort and the rocket bastion on it.

The assaulting columns were formed up ready at daybreak, 3rd April. The General's dispositions were, that a small detachment, under Major Gall, 14th Light Dragoons, should make a false attack on the western wall. As soon as the sound of his guns was heard, the storming parties were to debouch from cover and assault.

(*a*) The "Right Attack," composed of the Madras and Bombay Sappers, 3rd Bombay Europeans, and Hyderabad Contingent Infantry, was divided into two columns and a reserve. Its orders were to escalade the city wall near the Orchha Gate.

(*b*) The "Left Attack," in which were the Royal Engineers, the 86th Foot, and the 25th Bombay Infantry, was similarly divided into two columns and a reserve. The left column was to storm the breached Mamelon, while the right column escaladed the rocket tower and the low curtain, immediately to the right of it.

At 3 a.m. on 3rd April, the storming parties were ready in their respective positions, and no sooner was the signal given by Major Gall's party, than they dashed forward to the assault. Both columns of the "Left Attack," in spite of desperate opposition, were quite successful, and carried the Mamelon and city wall to the right of it. The "Right Attack" were not so fortunate; they had to force their way up the road to the Orchha Gate through a perfect hail of bullets, and when they did finally reach the wall some of their escalading ladders were found to be too short. Suffering heavy losses, the stormers, however, pushed on, and at length gained a footing on the wall, which the rebels were still fiercely contesting, when a party from the victorious "Left Attack" came to the aid of their comrades, by taking the rebels in flank and rear. Being thus succoured, the "Right Attack" was able to join the "Left," and take part in clearing the road to the palace, which is situated about four hundred yards from the out-works of the fort.

Throughout the entire length of this narrow city street, the British troops were obliged to fight their way, and here the struggle was terrible in the extreme, for it was for the most part a house-to-house and hand-to-hand encounter with desperate men at bay. As the conflict raged, the street was strewed with the bodies of the dead and dying; and the flames of burning houses, intensifying the heat of an April sun, made the temperature in the narrow road well-nigh unendurable. As it approached the palace the road ran close under the fort, with no houses to afford any cover to that side; consequently, the British troops were exposed to a heavy musketry fire from the crowds of the enemy on the fortress wall, which towered above them. The result was a number of casualties, which included several officers.

At the palace occurred the most sanguinary conflict of the day. The residence of the Rani had been specially prepared for resistance in the last resort. The courtyard was the scene of the first bloody encounter. Access to each apartment in turn was most stubbornly opposed, and to dislodge the rebels the bayonet was freely used. At length the struggle seemed to be at an end, but later on it was discovered that fifty men of the Rani's body-guard still held the stables attached to the palace. These to the last man stood their ground fighting to the death.

The struggle at the palace had but just terminated, when Sir Hugh Rose, who had been present throughout with the "Left Attack," received a report from the officer commanding one of the Hyderabad Contingent Cavalry camps, that a large body of the enemy, flying from the town, had tried to force his picquet. A few succeeded, but the Hyderabad Contingent Cavalry camps had, under the system of mutually supporting each other, rallied and driven back the main body, from 350 to 500 strong, on to a high, rocky, isolated hill to the west of the fort, near the Gwalior road, and there surrounded them till the arrival of reinforcements. The General at once ordered the available men of all arms, that had remained in camp, to move against the hill. The 24th Bombay Infantry led the attack, and the rebels died to a man. The Rani's father was wounded here, subsequently captured, and hanged at the place of the treacherous and cowardly massacre of the previous year. The lofty mass of rock, where the rebels made their last desperate stand against the bayonets of the 24th, and surrounded by the Contingent Cavalry, is to this day known as "Retribution Hill."

Desultory fighting continued in the city throughout the night, and early next morning, 4th April, it was discovered that the Rani had escaped during the night, and that the fort was abandoned. From information received, it appeared that the Rani, accompanied by 310 Wilayatis and 25 sowars, after leaving the fort had been headed back by one of the picquets. The party then separated, and the Rani, taking only a few sowars, turned off to the right towards Bandin. The observatory also telegraphed: "Enemy escaping to north-east."

The General immediately sent off strong detachments of the 14th Light Dragoons, 3rd Bombay Light Cavalry, and Contingent Cavalry, to pursue, with guns to support them, as it was said Tantia Topee had sent a force to meet her. Some cavalry were also sent to watch the fords of the Betwa. One of the pursuing parties, the 1st Cavalry under Lieutenant Dowker, found in the town of Bandin, over 20 miles from Jhansi, traces of the Rani's hasty flight, and her tent, in which was an unfinished breakfast.

On the other side of the town, he came up with and cut up 40 of the enemy, consisting of Rohillas and Bengal Irregular Cavalry. Lieutenant Dowker was gaining fast on the Rani, who with four attendants was seen escaping on a grey horse, when he was dismounted by a severe wound, and obliged to give up the pursuit. How extraordinary an achievement the Rani's escape was, will be better understood, when it is explained that the fort gate was on the side towards, and within a few yards of the palace and the Mamelon now in the hands of the British, and inside the city wall; also a considerable circuit had to be traversed by the fugitives before they reached the city gate—the Ganpat Khirki—through which they made their exit. Outside they then had to face the cordon of picquets.

During 4th April, the rest of the city was occupied, after some hard fighting, by a combined movement in which the cavalry co-operated from the north. From the time, however, of the capture of the palace, the rebels lost heart, and began to leave the town and fort. Nothing could prove more the efficiency of the investment, than the number of them cut up by the picquets of the flying camps. The woods, gardens, and roads, round the town, were strewed with the corpses of the fugitive rebels. The Rani's flight was the signal for a general retreat.

Early in the morning the outskirts of the city were scoured by cavalry and infantry, and it will give some idea of the destruction of insurgents which ensued, when it is related that a party of the 14th Light Dragoons alone killed two hundred in one patrol. The rebels generally sold their lives as dearly as they could, fighting to the last. A band of forty desperadoes barricaded themselves in a spacious house with a courtyard and vaults. This was attacked by a detachment of Hyderabad Contingent Infantry, under Captain Hare, and Captain Sinclair of the 3rd Contingent Infantry; the latter was killed in this affair. Reinforcements, including some guns, had to be brought up by Major Orr, but even when the house had been breached and knocked to pieces, the rebels continued to resist in the ruined passages and vaults till all were killed. Captain Abbott spoke highly of the gallantry with which Lieutenant Dun, 4th Cavalry, with dismounted detachments of the 1st and 4th Cavalry, stormed a house and garden obstinately held by fugitives.

This terminated the siege of Jhansi.

Sir Hugh Rose, in his despatch, writes:—

"The troops had to contend against an enemy more than double their numbers, behind formidable fortifications, who defended themselves afterwards from house to house in a spacious city, often under the fire of the fort, and later in the suburbs, and in very difficult ground outside the walls. The investing cavalry force were day and night for 17 days on arduous duty, the men not taking their clothes off, the horses saddled and bridled up at night. The nature of the defence and strictness of the investment gave rise to continued and fierce combats; for the rebels, having no hope, sought to sell their lives as dearly as possible. But the discipline and gallant spirit of the troops enabled them to overcome difficulties and opposition of every sort, to take the fortified city of Jhansi by storm, subduing the strongest fortress in Central India, and killing 5,000 of its rebel garrison. . . ."

Of the Hyderabad Contingent, Major Orr, Captains Abbott and Hare, and Lieutenants Dowker and Dun were mentioned in the despatch, as well as 13 native officers and men.

The total casualties of the British force during the operations against Jhansi, including the action with Tantia Topee, amounted to three hundred and forty-three killed and wounded, of whom thirty-six were officers.

Besides their losses at the battle of the Betwa, the 4th Cavalry had Captain Murray wounded, one duffadar and two troopers killed; two jemadars, the trumpet-major, and four troopers wounded at Jhansi.

Facing Page 104.

RETRIBUTION HILL,
JHANSI.

Rewards for conspicuous gallantry at Jhansi were:

4th Cavalry.

Trooper Bhagwan Singh, promoted duffadar.
Trooper Khairulla Khan, promoted duffadar.
Trooper Khari Md. Khan, promoted duffadar.
Trooper Yakub Khan, promoted duffadar.
Jemadar Hanuman Singh, 3rd Class Order of Merit.
Jemadar Mir Nur Ali, 3rd Class Order of Merit.
Duffadar Himmat Khan, 3rd Class Order of Merit.
Duffadar Bhagwan Singh, 3rd Class Order of Merit.

Later, Troopers Khan Muhammad Khan and Tabul Khan were also promoted duffadars for distinguished gallantry at this siege.

After the fall of Jhansi, Sir Hugh Rose remained there nearly nineteen days, partly to rest his troops and partly to make preparations for an advance on Kalpi, where Rao Sahib, a nephew of Nana Sahib, was; also an arsenal of artillery and warlike stores, and a numerous garrison. The Rani of Jhansi had fled there, and also Tantia Topee. The Rao Sahib's troops consisted of a number of regiments of the mutinied Gwalior Contingent, several battalions of regular Sepoys, all formerly belonging to the British, the contingents of various rebel Rajas, including the Nawab of Banda, and the remnant of the Jhansi garrison.

Sir Hugh Rose's first move immediately after the fall of Jhansi, was to send Major Orr with some of the Hyderabad Contingent Field Force, including the 1st and 4th Cavalry, across the Betwa to Mau, a place forty miles from Jhansi and beyond Barwa Sagar, where the rebels were said to have re-assembled. Thence the force was to proceed northwards to Gurserai, the Chief of which district was loyal, gain all possible information there, and then move against Kotra, an important ford across the Betwa, reported to be held by the rebels.

On the night of the 22nd April a column, under Major (now Brevet-Lieutenant-Colonel) Gall, consisting of one squadron 14th Light Dragoons, 3rd Cavalry Hyderabad Contingent, and three guns, was sent along the road to Kalpi, to gain information of the enemy. This column advanced to Punch, a place about 14 miles from Kunch; reconnoitring towards the latter place, the enemy were found to be in force there and preparing to resist the British advance. Here Major Orr, who had accomplished his object and marched to Erich, a ford across the Betwa, established communication between the two columns.

Leaving a small garrison at Jhansi, Sir Hugh Rose marched at midnight with the 1st Brigade, directing the 2nd Brigade to follow two days later.

The hot weather had now fully set in, and the troops suffered severely in consequence. The country they marched through was flat and without vegetation, and the dust lay several inches thick on the roads. The heat during the day obliged them to march at night, and many men died. The farther the force marched the scarcer became the water, now only found in wells at a very great depth, and then lukewarm and often brackish.

The object of the rebel leaders in sending their troops forward to Kunch, a distance of 42 miles, was to harass the British force, and if possible wear out at least the European portion of it, for it was known to be in a very exhausted condition from fighting in the sun. To render their opposition as effective as possible, the rebels decided to make no attacks on the British before ten o'clock in the day, in order to make them succumb more readily to the heat, and consequently either die or be sent to hospital.

The 1st Brigade reached Kunch 1st May. On the following day Major Gall was sent with a mixed force to Lohari, to exact reparation for an act of treachery. Some days previously he had posted a jemadar's party of the 3rd Cavalry, for the purpose of observing the enemy at that place, which was garrisoned by 70 or 80 men of the Raja of Sampta, who was believed to be an ally. These men, however, betrayed the Contingent Cavalry in the basest manner to the rebel cavalry at Kunch, but the former managed with some difficulty to cut their way out, with the loss of one man killed, some camp followers, and all their baggage. It subsequently turned out that the garrison were disguised Sepoys of the 12th Bengal Infantry, the regiment that had mutinied at Jhansi.

Major Gall marched at 2 a.m., invested the place

with cavalry, blew in the gate, and stormed the fort with infantry. The garrison fought desperately, but to the number of about 90 were killed to a man.

On 5th May Sir Hugh Rose was joined by the 2nd Brigade, now reinforced by the 71st Highland Light Infantry.

Meantime Major Orr and his party of the Hyderabad Contingent, including the 4th Cavalry, had been ordered to try and prevent the Rajas of Banpur and Shahgarh and other bodies of rebels crossing the Betwa, and doubling back southwards. The two Rajas, for the purpose of carrying out this very manœuvre, had separated from the rebels at Kunch, and driven the troops of the Raja of Gurserai, who held Kotra, commanding one of the fords across the Betwa, to the south bank of the river.

Major Orr engaged the Rajas, drove them from their position at Kotra, and took one of their guns; but he found it impossible to cut off their retreat southwards, as, while he was engaging one portion, the remainder retired precipitately some distance down the river and crossed by another ford.

Major Orr then marched on Kunch, before which he arrived on the morning of the battle and in time to take part in it.

The entire force marched at 10 p.m., 6th May, with a view to turning the enemy's flank, threatening his line of retreat on Kalpi, and attacking Kunch from the north-west. After a march of fourteen miles a position was reached two miles from the town, and facing its unfortified side.

When day broke on 7th May all was made ready for the attack. The heat was excessive, and there were 46 cases of sunstroke. The diary of an officer present gives some interesting details. He writes: "The heat at daybreak was intense and the mirage most remarkable; the whole of the surrounding country was dried up, covered with light brown soil, and perfectly flat, yet it appeared one beautiful lake of water, and the few trees assumed the appearance of gigantic height. When Major Orr and the Hyderabad Contingent Cavalry approached to rejoin our force, so distorted were they that we could not tell whether they were friends or foes. The horses appeared twenty feet high, and the riders gigantic in proportion, and the heated air ascending made the figures tremulous and crooked."

Sir Hugh Rose suffered much from the sun this day, and was obliged to dismount for a time, and seek shade and medical assistance.

The British force was formed for the attack with the 1st Brigade on the left at Nagapura; the 2nd Brigade, in the centre, occupied the village of Chumer; whilst Major Orr, with the Hyderabad Contingent was on the right at Umri.

At 8 a.m. Major Gall was sent forward to reconnoitre the gardens, woods, and temples which lay between Kunch and the British Army. His advance was covered by artillery fire, and at the same time the siege guns were moved into a position whence they could effectually play on the town.

Major Gall reported that the enemy had retreated through the woods nearer the town, that they had cavalry in their rear, that the fire of the siege guns had driven the rebels from the right of the wood into the town, but that some earthworks were still held by them.

The 1st Brigade then advanced, cleared the wood and earthworks, and seized the fort, though the fighting and skirmishing in the wood caused several casualties. The enemy were ultimately driven through the town and along the plain on the road leading towards Kalpi. This was on the north side. The 2nd Brigade, advancing from the west, attacked some strongly-posted rebel infantry to their front, meeting with a very determined resistance; the 1st Brigade, however, assisted with a well directed flank attack, and the rebels were dispersed. In the latter attack two troops of the 14th Light Dragoons, supported by a troop of the 3rd Cavalry, made a fine charge against a mass of the enemy, which they broke and cut up. On the right, Major Orr with the Hyderabad Contingent, moved direct on the town from Umri, clearing the rebels out of the gardens and enclosures. A large body of the enemy's cavalry then threatened the right flank, but were charged by the 1st and 4th Cavalry, dispersed, and driven off the field. These two regiments then joined in the cavalry pursuit.

The enemy now retreated up the Kalpi road, but without hurry or disorder, and the whole of the cavalry, horse artillery and light guns took up the pursuit.

The 14th Light Dragoons made some fine charges, cutting up the enemy's rear guard. Captain Blyth, of the 14th, and Captain Abbott each captured a gun, charging under heavy fire. This pursuit was continued up to seven miles from Kunch, in the burning sun, and men and horses suffered intensely; it was 8 p.m. before they got back, having been 16 hours in the saddle. The infantry were too exhausted by the heat to assist, though the actual action was over in about an hour.

Between 500 and 600 of the rebels were killed, and nine guns and a quantity of ammunition taken.

The British loss amounted to three officers and 59 men killed and wounded, as well as many casualties from sunstroke.

In the Hyderabad Contingent Cavalry the casualties were:—

1st Cavalry	...	1 killed, 8 wounded.
3rd Cavalry	...	1 killed, 2 wounded.
4th Cavalry	...	1 killed, 5 wounded.

Among others, the services of the following Contingent Cavalry officers were brought to notice:

Captain Murray (4th), Lieutenants Dowker (1st), Dun (4th), Fraser (4th), Westmacott (4th), Johnson (1st), Surgeon Orr (4th), and Assistant Surgeon Sanderson (1st).

The Rani of Jhansi, who had been present during the action, fled to Kalpi; and Tantia Topee, always foremost in flight, escaped to a place near Jalaun, where his parents dwelt. The rebel Sepoys of the Bengal Army fought well, especially in the retreat, in which they showed considerable military training; the mutineers of the 52nd Bengal Infantry, who formed the rear guard, were almost annihilated. The rebel cavalry, on the other hand, distinguished themselves by their cowardice. The scorching rays of the sun and pace of the retreat told even on the rebels, several of whom fell dead on the road from heat apoplexy.

The 8th May was a day of rest for both brigades, and at 2 a.m. on the 9th the General marched with the 1st Brigade, but was obliged to halt at Hardawi, the 2nd Brigade, which had received orders to follow the next day, having been delayed by a terrific storm.

The condition of the opposing forces now was briefly this: The rebels, after their succession of defeats, were absolutely discouraged and demoralized, and had become so distrustful of one another that, on hearing of the fresh advance of the British, the greater part dispersed in all directions. But for an unexpected event, which occurred a few days afterwards, there might have been no more fighting; this event was the arrival of the Nawab of Banda with a force of two thousand mutineer cavalrymen and some guns, besides his own followers. The rebel leaders now put forth their utmost endeavours, to render their preparations for defence as complete as possible, and the troops that had dispersed returned to Kalpi. All this, as Colonel Malleson remarks, "produced one of those changes from despair to confidence which mark the Indian character."

As regards the British force, the sun, now at its maximum heat*, was a formidable ally to the rebels;

* At Kunch the thermometer was 115 degrees in the shade, before Kalpi 118 degrees, and later on the march to Gwalior, it burst in an officer's tent at 130 degrees.

its blaze made havoc amongst the troops, especially the Europeans, already exhausted by months of over-fatigue, want of sleep, continuous night watching, and night marches, and often exposed to its rays, manœuvring or fighting, from sunrise to sunset. The 71st Highland Light Infantry suffered the most from this cause, having just landed in India, and not being inured. The number of officers and men on the sick list and carried in doolies increased daily, and there was no water except for a few wells. Forage was as scanty as water, and this crippled the cavalry and transport, while the enemy's cavalry were numerous and enterprising. The scarcity of water also prevented the concentration of the force. The inhabitants of the Jumna Valley were most hostile, and acted as spies to the enemy. Lastly, the Enfield rifles had deteriorated, owing to heat and other causes. The ammunition had become in such a state that loading was difficult and fire uncertain; the men lost confidence in their arms.

Kalpi was the best rebel fortified stronghold in Central and Western India; as a fortification it was insignificant, but as a position unusually strong, being protected on all sides by interminable ravines, to the front by five lines of defence, and to its rear by the Jumna, from which rises the precipitous rock on which the fort stands. Malleson describes the lines of defence as:—

"1st.—A series of entrenchments, with flank defence.

"2nd.—Eighty-four temples of solid masonry, with walls round them of the same.

"3rd.—An outwork of ravines.

"4th.—City of Kalpi.

"5th.—A second chain of ravines.

And lastly came the fort itself."

The rebels, with some reason, considered the position almost impregnable, but they had no other idea, than that the British would make their attack from the direction of their approach from Kunch.

This, then, was the situation which Sir Hugh Rose had to face, and he realised how much depended upon the issue of the battle to be fought. Anticipating that the utmost desperation would be shown by the rebels, in what they would regard as their last encounter with the British, he determined that nothing should be lacking in his preparations for striking the blow, which he intended should end a campaign, which was fast wearing out not only himself, but all under his command.

He decided not to attempt a frontal attack, but to turn to the right from the high road to Orai and Kalpi, and march to the Jumna, at the village of Golauli, five miles down stream from Kalpi, and there establish communication with Colonel Maxwell, commanding a column of the Bengal Army, who had been sent to co-operate with him and bring fresh supplies of ammunition. He could then move up the Jumna on Kalpi, covered by the flank fire of Colonel Maxwell from the other side of the river. Another advantage from this plan was, that he could obtain water from the river; elsewhere it was very scarce.

To mislead the enemy and mask this movement, he ordered the 2nd Brigade to move direct on Orai; unfortunately, however, this brigade missed its way, and had to make a double march in the sun, which caused many casualties and much exhausted them.

The two brigades concentrated on the night of the 14th May at Itaura. A few hours later Sir Hugh Rose, with the 1st Brigade and the Hyderabad Contingent Field Force, marched for Golauli. They crossed the high road from Jalalpur to Kalpi, and Major Orr and the Hyderabad Contingent were directed to drive back the enemy's picquets on that road, and remain in position there to cover the march of the rear guard. Having done this, they then encamped at the adjacent village of Tehri, to watch the road, keep up communication with the 2nd Brigade in rear, and assist it on its march during the 15th to Deopura, another village close by.

The General, with the 1st Brigade, reached Golauli with very little opposition, and on arrival there two men of the Hyderabad Contingent Cavalry were despatched across the river to Lieutenant-Colonel Maxwell, who was thirty miles distant, to direct him to move up the river immediately. These men reached their destination in safety. At the same time two pontoon rafts, which had been brought from Poona, were floated on the Jumna to establish communication with the other bank, the rebels having destroyed or removed all the boats.

The advanced guard and main body of the 2nd Brigade reached Deopura practically unopposed, but the rear guard, commanded by Major Forbes, C.B., which had been reinforced by 200 sabres of the 1st and 4th Cavalry, Hyderabad Contingent, under Captain Murray, had hardly left Itaura when it was vigorously attacked by between a thousand and twelve hundred rebel cavalry, supported by three or four thousand infantry, with heavy guns drawn by elephants. The 4th Cavalry made repeated charges, but the baggage became entangled in a broad, deep ravine, only passable by one cart at a time, and the situation was only saved by the cool, steady front shown by the rear guard, now reinforced by the remainder of the Hyderabad Contingent, under Major Orr. For the first three miles after extricating the baggage from the ravine, the rear guard and Hyderabad Contingent were almost surrounded by rebel cavalry, and under artillery fire; but alternately halting and retiring, they succeeded in preserving the baggage, which was brought in safety into Deopura. Sir Hugh Rose hearing the rear guard were hard pressed, moved rapidly from Golauli to Deopura with a force of all arms, but on arrival there found the whole lot safely in.

The baffled enemy, reinforced by several fresh battalions, now attempted to storm the village of Mutha, close by. This village dominated the camp at Deopura, and was held by a force under Lieutenant-Colonel Campbell, who, finding himself hard pressed, had just given orders to evacuate it, when Sir Hugh opportunely arrived. The retirement was at once stopped, and reinforcements ordered up, including half a Hyderabad Contingent battery, the 3rd Cavalry, and two companies of Hyderabad Contingent Infantry. The General, concentrating a heavy artillery and infantry fire on the rebels, drove them back, with heavy loss, into Kalpi, but did not pursue, as the sun was already causing frightful havoc among his men.

The British force was now disposed with the 1st Brigade on the Jumna at Golauli, the Hyderabad Contingent at Tehri, and the 2nd Brigade at Deopura and Mutha. Along the whole of this line the enemy maintained a succession of futile attacks till sunset, when they withdrew. The following afternoon they again attacked Deopura, and were repulsed with loss.

Meanwhile, Colonel Maxwell arrived on the other side of the river, 18th May, and the same night commenced constructing batteries to shell the fort of Kalpi, blow up, if possible, the powder magazines in it, and destroy the defences facing the British position at Golauli. Another mortar battery was placed lower down the Jumna, opposite the village of Rayar, an important point on the line of advance from Golauli, where the rebels had posted a force and a battery.

On the morning of the 19th, the 2nd Brigade and Hyderabad Contingent were brought in to camp on the Jumna. Water had failed in the villages, the troops had suffered severely from the heat, and were being harassed incessantly by attacks.

On the night of the 20th, Colonel Maxwell sent across the river reinforcements of two companies of the 88th Regiment, a camel corps, and some Sikh infantry; later on some guns also were sent.

On 21st May, the General, having learned that the enemy intended to bring on a battle the next day, resolved to deliver without further delay the blow, which he now felt he could safely strike. He describes his dispositions as follows:—

"The right flank, facing the left of Kalpi, rested on the ravines running down to the Jumna; in these ravines were the villages of Salauli and Golauli. Both these villages were held and protected by strong picquets, and prevented my right being turned.

"Half of the 1st Brigade, my right flank, was encamped perpendicularly to the Jumna, facing the belt of ravines, and the left front of Kalpi, on the table land, immediately outside the belt.

"The remainder of the 1st Brigade facing the continuation of the belt of ravines, which took a sweep outward, and the 2nd Brigade and Hyderabad Field Force, facing the table land or plain stretching from Kalpi to Jalalpur, were thrown back *en potence*. This ground was adapted to the movements of artillery and cavalry.

"My whole front was guarded by strong outposts with advanced sentries in the ravines, and picquets."

Meantime, the rebels prepared for the attack by swearing a solemn oath on the sacred waters of the Jumna, that they would drive the British Force into the river or die, and by issuing large quantities of opium to the troops to nerve them for the battle.

On the morning of the 22nd May, Sir Hugh Rose made his final dispositions, in which the Hyderabad Contingent were placed on the extreme left, with Maxwell's Camel Corps Riflemen and Sikhs.

Shortly after 8 a.m. the enemy advanced in great force towards the belt of ravines on the British right, and along the Jalalpur-Kalpi road against the left, manœuvring so skilfully that they got unperceived into the ravines on the right, where their real attack was carried out. Meantime, the enemy's right, consisting of about 1,400 cavalry, with several battalions of infantry and some 9-pounder guns, continuing their advance, threatened to outflank the British left.

Sir Hugh Rose was not, however, deceived by these manœuvres, and still regarded his right as the threatened flank. He, however, reinforced his picquets on the left with a squadron of the 14th Light Dragoons and the 3rd Cavalry, under Captain Abbott, and afterwards directed these troops to retire obliquely across his front, in order to conceal his heavy guns and draw the enemy into their fire. This manœuvre was partially successful; the enemy's cavalry lost heavily from the fire of the guns.

The General's forecast was now proved correct. Suddenly, as if by magic, the whole line of ravines became lighted up by a mass of fire, both artillery and musketry, which was brought to bear with overwhelming force on the British right. As Malleson says: "The suddenness of the attack, the superior numbers of those making it, and the terrible heat of the day, gave the rebels a great advantage." The sun also had struck down numbers of men, so that things looked very critical, when the enemy, maddened with opium, swarming from the ravines, pressed forward, and caused the British to fall back to the position where the light field guns and mortar batteries were posted. Brigadier Stuart called on the gunners to defend their guns with their lives, while the 86th Regiment and 25th Bombay Infantry step by step disputed the enemy's advance. Sir Hugh Rose, who was watching the progress of the battle on the left, observed a slackening in the British fire on the right and an increase in that of the enemy. Taking with him the Camel Corps, he hurried to that flank, and, dismounting the men there, led them himself forward at the double, and charged the advancing foe, then within a few yards of the British guns, which had ceased firing. For a moment the enemy stood, but only for a moment. There was a shout and a dash forward from the whole British line, and the rebels were driven headlong into the ravines below. Not only was the attack on the right repulsed, but the battle was also won. The attack on the left collapsed when it was seen that the one on the right had failed, and the guns, gaining the rebels' flank, inflicted great loss. The whole of the cavalry made a converging movement against the enemy's right and village of Tehri. The rebels then broke and fled, pursued for some miles by the cavalry and horse artillery; numbers were killed, and they were so closely followed up that many were cut off from Kalpi. They were completely beaten and disheartened, despite their having displayed unprecedented energy and skilful tactics. Meantime, Maxwell's batteries had been bombarding the fort and town to such effect that the enemy found it no secure place of refuge. They evacuated it accordingly.

The General decided that an immediate advance on Kalpi would be successful, and marched early next morning, 23rd May, in two columns of attack, while Colonel Maxwell's batteries co-operated by shelling the fort and villages in the line of advance.

The right column, composed of the 1st Brigade, followed the course of the Jumna, and went through the ravines. The left column, or 2nd Brigade, under the personal command of Sir Hugh Rose, advanced along the Jalalpur-Kalpi road, covered by the Camel Corps, supported by the Hyderabad Contingent Cavalry. Neither column met with any strong resistance, and by 10 a.m. the town and fort of Kalpi were occupied, having been deserted by the mutineers; guns, standards and every description of warlike stores were taken.

Large bodies of the enemy, with elephants, were reported to be retreating by the Jalaun or Gwalior road, and Lieutenant-Colonel Gall, with four troops of the 14th Light Dragoons, six guns horse artillery, Captain Abbott, with the 3rd Cavalry, and Lieutenant Dowker, with 50 sabres of the 1st Cavalry, were ordered in pursuit. The rebels were caught retiring across a plain, charged, dispersed, and many sabred. Several guns and waggons were captured, also six elephants. Lieutenant-Colonel Gall, in his despatch on this pursuit, particularly brought to notice the gallant way Captain Abbott led his men; he also mentioned Lieutenant Dowker.

The losses of the Hyderabad Contingent Cavalry in the operations round Kalpi were:—

1st Cavalry.—One trooper killed and one wounded.

3rd Cavalry.—One duffadar wounded, one trooper killed, and one wounded.

4th Cavalry.—Jemadar Chotay Khan and three troopers killed, two duffadars and one trooper wounded.

In his despatch reporting the operations before Kalpi, Sir Hugh Rose said:—"So much of the success of the operations is due to the portion of the Hyderabad Contingent, which formed part of my force, that I ought not to fail to express my best thanks to Colonel Davidson, Resident at Hyderabad, for the proof of confidence which he placed in me, by putting at my disposal troops, whose organisation in the three arms, light equipment, knowledge of the Indian language and country, combined with their high military qualities, enabled them to act as the wings of my operations."

Of what had been accomplished at Kalpi on that eventful day, Dr. Lowe, an eye-witness, wrote thus:—"A glorious victory was won over ten times our numbers, under most trying circumstances. The position of Kalpi; the numbers of the enemy, who came on with a resolution and a display of tactics we had never before witnessed; the exhausted, weakened state of the General's force; the awful, suffocating hot winds and burning sun, which the men had to endure all day without time to take food or water, combined to render the achievement one of unsurpassed difficulty."

Major Orr brought to notice the following officers of the Hyderabad Contingent Cavalry with him for their intelligence and devotion to duty during the operations round Kalpi:—

Captain Murray, Commanding 4th Cavalry.
Lieutenant Dowker, Commanding 1st Cavalry.
Lieutenant Dun, Second-in-Command, 4th Cavalry.
Lieutenant Fraser, Adjutant, 4th Cavalry, and Staff Officer, Field Force.
Lieutenant Westmacott, 4th Cavalry.
Lieutenant Johnson, Adjutant, 1st Cavalry.
Surgeon Orr, 4th Cavalry.
Assistant-Surgeon Sanderson, 1st Cavalry

The Rani of Jhansi and Nawab of Banda had both made a precipitate retreat from Kalpi, at midnight, after the battle. It was said that a shell from Maxwell's batteries burst in the Rani's room, killing two of her attendants, which hastened her departure. They first went to Gopalpore, a small town in the jungle, forty-six miles south-west of Gwalior. There Tantia Topee joined them. The fresh schemes which they then evolved will be seen shortly.

On 25th May, Lieutenant-Colonel Robertson, with one troop, 14th Dragoons, one squadron 3rd Bombay Cavalry, 160 sabres Hyderabad Contingent Cavalry, and a light field battery was sent to follow up the enemy along the Jalaun road. On 29th May this column was reinforced from Kalpi by two squadrons of the Fourteenth and a wing of the 86th Regiment, and, pushing forward shortly after, reported that there was no doubt that the main body of the rebels had taken the road to Gwalior.

In the meantime Major Orr, who was at Orai with the Hyderabad Contingent Field Force, on 30th May received information that a noted rebel, Thakur Barjur Singh, was at his stronghold, Bilayan, 19 miles off. He decided to move at once, and by making a forced march surprise and surround him.

He accordingly marched at 10 p.m. with—

- 1st Cavalry, 165 sabres, Lieutenant Dowker Commanding.
- 4th Cavalry, 162 sabres, Lieutenant Dun Commanding.
- Six guns of the Hyderabad Contingent Artillery.
- 3rd Infantry, Hyderabad Contingent, 344 bayonets, Lieutenant Macquoid Commanding.
- Siege train, one 18-pounder gun, and two 8-inch mortars.
- Sappers and miners, 22.

Taking the cavalry with him, Major Orr pushed on to Bilayan, and, arriving at daybreak, surrounded the place, and waited for the infantry and guns. The enemy opened fire, and then from 200 to 250 sallied out in a compact body. Leaving picquets to watch the fort, Orr collected 81 men of his cavalry and advanced against this mass, to embarrass and impede their progress till the artillery and infantry, which were by now in sight, could come up.

The guns opened fire with shell and shrapnel with good effect. The cavalry then charged and cut up a number of both horsemen and foot; but the remainder escaped into some deep and strong ravines, running towards the Betwa, to attack or dislodge them from which was very difficult, as the place was a mass of prickly thorn bushes. This was, however, accomplished by the 3rd Infantry, after some desperate fighting, in which they lost seven killed and one subadar and three men wounded.

Barjur Singh, unfortunately, escaped, but his horse and standard were taken. He himself only got away by threading the ravines on foot, having stripped himself of almost all his clothing. He lost everything he had. His horses, camels, and all his property were captured, and his power and influence completely broken by the death or dispersion of his entire band.

It was estimated that between 120 and 130 rebels were killed here; 35 besides were made prisoners.

Major Orr brought to notice in his despatch, for gallantry, alacrity, and zeal, Lieutenants Dowker (1st), Macquoid (3rd Infantry), Dun (4th), Fraser (4th), and Westmacott (4th), Surgeon Orr (4th), and Assistant-Surgeon Sanderson (1st), as well as some native officers and men.

1st Cavalry lost one duffadar and one trooper wounded. The 4th Cavalry had Lieutenant Westmacott severely, and one trooper slightly, wounded.

In his despatch forwarding Major Orr's report on these operations, Sir Hugh Rose wrote:—"Major Orr, in crushing rapidly and effectively this rebel, did essential service to the Government and my Force, for which I beg to recommend him strongly to His Excellency, submitting to his favourable notice at the same time, the officers named in his despatch, as well as the native officers and men who have been recommended to me by him for distinguished conduct."

It was thought that, with the fall of Kalpi, the campaign in Central India had come to a close. The enemy had been defeated wherever met with, and his forces were scattered in every direction. Sir Hugh Rose even issued the following farewell order to the troops:—

"Soldiers! You have marched more than a thousand miles, and captured more than a hundred guns. You have forced your way through mountain passes and intricate jungles, and over rivers. You have captured the strongest forts and beat the enemy, no matter what the odds, wherever you have met him. You have restored extensive districts to the Government, and peace and order now, where before for a twelvemonth were tyranny and rebellion. You have done all this, and you have never had a check. I thank you with all my sincerity for your bravery, your devotion, and your discipline."

* * * * * * * *

The Commander-in-Chief in India issued a General Order to the various forces in the field on the 28th May, 1858, from which the following extracts are taken:—

* * * * * * * *

"the three columns put in movement from Madras and Bombay have rendered likewise great and efficient services in their long and difficult marches to the Jumna, through Central India, and in Rajputana.

"These columns, under the command of Major-Generals Sir Hugh Rose, K.C.B., Whitlock, and Roberts, have admirably performed their share in the general combination arranged under the orders of His Lordship the Governor-General.

"That combination was spread over a surface ranging from the boundaries of Bombay and Madras, to the extreme North-West of India.

"By their patient endurance of fatigue, their unfailing obedience, and their steadfast gallantry, the troops have enabled the Generals to fulfil their instructions.

"In no war has it ever happened, that troops have been more often engaged than during the campaigns, which have now terminated.

"In no war has it ever happened, that troops should always contend against immense numerical odds, as has been invariably the case in every encounter during the struggle of the last year, and in no war has constant success without a check been more conspicuously achieved.

"It has not occurred that one column here, or another there, has won more honour than the other portions of the Army.

"The various corps have done like hard work,

have struggled through the difficulties of a hot-weather campaign, and have compensated for paucity of numbers in the vast area of operations by continuous and unexampled marching, notwithstanding the season.

"It is probable that much yet remains for the Army to perform, but now that the Commander-in-Chief is able to give the greater part of it rest for a time, he chooses this moment to congratulate the Generals and the troops on the great results which have attended their labours.

"He can fairly say that they have accomplished in a few months what was believed by the ill-wishers of England to be either beyond her strength or to be the work of many years."

* * * * * * * *

The 4th Cavalry now started, with the rest of the Hyderabad Contingent Field Force, to march back to the Deccan, but did not get far before a fresh outburst of the flames of rebellion necessitated their return.

The Rani of Jhansi, desperate and daring, conceived the plan of marching on Gwalior and taking possession of that stronghold. On 30th May Tantia Topee, the Rani, and other leaders, with 7,000 infantry, 4,000 cavalry, and 12 guns, entered Morar, the cantonment of the mutinied Gwalior Contingent, situated a little east of the fortress of Gwalior. At daybreak next morning, 31st May, the Maharajah Sindhia marched out, with some 8,000 men and eight guns, and took up a position at Bahadurpur, two miles east of Morar.

The rebels advanced at 7 a.m., carried the guns by a charge of 2,000 cavalry, and then the whole of Sindhia's army, except his bodyguard, went over to them. The bodyguard made a gallant defence, but was overwhelmed, and Sindhia fled to Agra.

The rebels then entered Gwalior, and seized the treasury and Maharajah's jewels, the latter said to be of fabulous value. The garrison of the fortress opened its gates, and 50 to 60 guns, and a fine arsenal, stocked with warlike stores, fell into the mutineers' hands.

The troops that went over were the best drilled and organised of all the native levies.

The time of year was the most unfavourable for military operations, it being the eve of the rainy season, and the summer heat at its maximum. It was absolutely essential that the troops should reach Gwalior before the rains set in, as there were no pontoons for the siege artillery, and it would be impossible to transport it across the Pahuj and Sind Rivers, when in flood.

Sir Hugh Rose received information on 3rd June from Colonel Robertson's column, which, as above mentioned, had left Kalpi to follow up the retreating rebels, that Gwalior had fallen. He had suffered so much in health that he had given up his command, but at once resumed it, and decided to march on Gwalior immediately. On first news of the new developments, Brigadier Stuart had been sent with a force of all arms to reinforce Colonel Robertson. Leaving a garrison at Kalpi, the General himself followed on 6th June, taking with him a troop of horse artillery, one squadron 14th

Light Dragoons, one squadron 3rd Bombay Cavalry, and Madras Sappers and Miners. Proceeding by forced marches at night to avoid the sun, he came up with Stuart at Indurkhi, on the Sind River, on 11th June. Here information was received that a column under Colonel Riddell, composed of the 3rd Bengal Europeans, 200 Sikh Horse, 300 Sikh Infantry, a light field battery, and some siege artillery, were marching from Agra to Gwalior, to co-operate with him; while Brigadier Smith, with a brigade of the Rajputana Field Force, consisting of 8th Hussars, 1st Bombay Cavalry, 95th Regiment, and 10th Bombay Infantry, were also moving on Gwalior for the same purpose. To complete matters the Hyderabad Contingent Field Force, on hearing the news, at once turned back, and marched on Gwalior by way of Jhansi. Regarding the last-named, Sir Hugh Rose wrote in his despatch:—"The Hyderabad Contingent, after their hard service, had received permission and orders to return home; almost all these troops had commenced their return to the Deccan, and some of them were far advanced on their road. With a good feeling, which cannot be sufficiently praised, all of the Contingent, which formed part of the Central India Field Force, instantly counter-marched and moved towards Gwalior, on the wish being intimated to the Officers Commanding their separate bodies that they should perform this fresh act of good service for the Government."

Agra was selected as the base of operations, its communications with Gwalior being short and good, though rendered imperfect by the passage of a very difficult ford across the Chambal River.

Sir Hugh Rose decided to invest Gwalior as much as its great extent would allow, then attack the weakest side, the investing troops cutting off the escape of the rebels. In accordance with this decision, he directed Major Orr, with some of the Hyderabad Contingent Field Force, who had joined Brigadier Smith's column on the road from Jhansi to Gwalior, to move to Punniar, on the road from Sipri to Gwalior. Here, though too weak to attack, he was perfectly placed for cutting off the retreat of the rebels towards the south. From the very meagre accounts available, it would appear that the 4th Cavalry accompanied Major Orr during the subsequent operations.

Brigadier Smith was ordered to march to Kotah-ki-Serai, about seven miles south-east of Gwalior, and separated from it by a mass of hills, covered with rocks and thick jungle.

Colonel Riddell, with the force from Agra, was ordered to advance to the Old Residency, about two miles on the northern side of Gwalior Fort.

Finally, he himself joined Brigadier Stuart's column, and marched against Morar cantonments, situated about five miles east of Gwalior, and which had been only partially burned by the rebels. His plan was to capture Morar and establish there, in the buildings still intact, his hospital, parks, etc. Divested of these encumbrances, and leaving a force there to protect them, form part of the investment, and pursue when required, he would be free to join

Brigadier Smith at Kotah-ki-Serai, and attack Gwalior thence with both brigades.

Sir Hugh Rose arrived 16th June at Bahadurpur, close to Morar and the scene of Sindhia's defeat, and sent Captain Abbott, with his Hyderabad Contingent Cavalry, which consisted of the 3rd Regiment and 50 men of the 1st, to reconnoitre. Abbott reported the enemy in force in front of the cantonments. The General, fearing that if he delayed till the next day the rebels might burn the remaining buildings, and also knowing that a prompt attack had always the greatest effect against them, countermanded the orders for encamping, and made his dispositions for an immediate attack.

Captain Abbott, with his cavalry, covered the advance, and came under fire of a masked battery, which caused some casualties, but, as Sir Hugh Rose himself remarked, "his men showed admirable steadiness." Ravines and broken ground prevented the farther advance of the cavalry. The infantry now attacked, and carried the place most gallantly, the enemy offering a desperate resistance, and finally retreating to Gwalior from their right. A wing of the 14th Light Dragoons made a very successful pursuit, catching the retreating rebels in the plain and slaughtering great numbers of them. Captain Abbott, whose horse had been killed by a round shot, contrived to get round the nullahs with his Contingent Cavalry, and ably assisted in the pursuit, as did also the "Eagle" troop of the Bombay Horse Artillery, who greatly distinguished themselves by their rapid and skilful advance.

On the morning of 17th June, Smith's Brigade of the Rajputana Field Force advanced from Antri to Kotah-ki-Serai, and found the enemy with guns occupying the heights between there and Gwalior. Deciding that it would not be advisable to camp in the plain with the enemy in position above him, Brigadier Smith attacked them with his infantry, drove them back, and then moved forward his cavalry. The road, before debouching from the hills, ran for several hundred yards in a defile, through which a squadron of the 8th Hussars pushed on down into the outskirts of Gwalior, swept through the enemy's camp, completely clearing it, taking some guns and cutting down the mutineers by scores. They then continued the charge right through the Phul Bagh cantonment, close under the southern end of the fort. Here the Rani of Jhansi, dressed as a cavalry soldier, was cut down by a Hussar, ignorant of her rank and sex. The squadron then withdrew by alternate troops, covered by their artillery, and supported by a troop of the 1st Bombay Lancers. The officers and men were so completely exhausted and prostrated from heat and fatigue, that they could scarcely sit in their saddles, and were for the moment incapable of further exertion. Smith determined to content himself with holding for the night the defile and the adjoining hills on the right. The enemy, who appeared to be threatening an attack, held their ground on the heights to the left. During the whole of the 18th these positions were maintained, the enemy keeping up a constant artillery fire.

Meantime, on the afternoon of the 18th, Sir Hugh Rose, leaving a force, chiefly cavalry and including the 3rd Contingent Cavalry and 50 men of the 1st, for the protection of Morar, the investment of Gwalior, and the pursuit of the enemy when they should retreat, marched from there for Kotah-ki-Serai, to support Brigadier Smith. The distance to be traversed was long and circuitous, and the heat terrible; one hundred men of a single regiment were struck down by the sun, and the troops reached Kotah-ki-Serai that night in a somewhat exhausted state after their twenty-mile march.

The following morning, 19th June, the British troops attacked and carried the heights above Gwalior to the left of the road, and followed this up by a general attack on Gwalior itself, and before sunset the whole of the Lashkar, or new city, was in their hands. The rebel cavalry and infantry retreated with great rapidity, but numbers were killed at the Phul Bagh, and a large number were pursued round the rock of the fort, their guns taken and many killed.

The fort, which is situated on the top of a long, flat-topped, isolated hill, rising precipitously on all sides from the plain, was closely invested, and the following morning daringly captured by Lieutenants Rose and Waller and a small detachment of the 25th Bombay Infantry. Accompanied by a blacksmith, they crept up, forced the gates, and surprised and killed the whole garrison. Unfortunately Lieutenant Rose, a young officer of brilliant promise, was among the killed.

Tantia Topee, with a large body of cavalry and infantry, attempted to retreat by Punniar, but was met and hunted back by Major Orr and the Hyderabad Contingent. He then went to the old Residency, where the rebels rallied. This place should have been occupied by Colonel Riddell's column, but they had been delayed at the ford over the Chambal River and had not arrived.

On the morning of the 20th orders were sent to *Brigadier Napier at Morar to pursue the enemy. Marching before 7 a.m., with detachments of—

	Officers & Men.
1st Troop, Horse Artillery	99
14th Light Dragoons	62
3rd Bombay Light Cavalry	104
3rd Contingent Cavalry	245
Meade's Horse	180
Total	690

he reached Samaoli, 25 miles distant, in the evening, and there learnt that the enemy had gone on to Jaora-Alipur, with 12,000 men and 22 guns. Marching at 4 a.m. on the 21st, he found the rebels strongly posted, with their right resting on Alipur, guns and infantry in the centre, and cavalry on both flanks.

Captain Abbott, with his cavalry, covered the advance, and the "Eagle" Troop of Horse Artillery were directed to take up a position about 600 yards from the enemy's left flank, enfilade them, and to act afterwards as circumstances might dictate.

* Afterwards Lord Napier of Magdala.

The guns opened and quickly silenced those of the enemy. The cavalry advanced at the gallop; then Captain Abbott charged. The movement was spontaneously followed by the rest of the cavalry and the guns, and they swept through the enemy's batteries and camp, driving the rebels before them for miles and cutting them down. Wherever a body of them collected, the horse artillery opened and dispersed them; it was said by an eye-witness that the dash of the gunners was simply wonderful.

The pursuit was carried on for six miles, and the enemy were dispersed in every direction, throwing away their arms. Twenty-five guns were captured, and between three and four hundred of the enemy killed. Some of the villages in rear were still full of the enemy, but Abbott dismounted some of his men with swords and carbines, and, aided by the fire of two guns, cleared the whole lot out. Besides the guns, quantities of ammunition, an elephant, tents, carts, and baggage were taken.

The British loss was only four killed and eight wounded; among these one killed and three wounded belonged to the 3rd Contingent Cavalry.

General Napier, in his despatch, writes:—"Captain Abbott, Commanding the Hyderabad Cavalry, distinguished himself greatly by his activity and intelligence generally, and by the gallantry of his charge on the enemy's batteries."

The whole of the Hyderabad Contingent Field Force now marched back to their stations in the Deccan.

The following Brigade Order, by Brigadier C. S. Stuart, C.B., Commanding 1st Brigade, Central

India Field Force, was published on the 28th June, 1858:—

"The 3rd Regiment Cavalry, Hyderabad Contingent, being about to leave this brigade, Brigadier Stuart considers he simply discharges an act of duty in adding his testimony to the worth of this distinguished regiment. For upwards of a year the men of this corps have been unceasingly engaged in most arduous and trying duties; they have ever shown the greatest cheerfulness, and on every occasion of their coming in contact with the enemy their gallant conduct and marked success have been conspicuous. The Regiment leaves this brigade carrying with it the sincere regret and esteem of its late comrades of all ranks; greater praise than this the Brigadier considers would be difficult to convey to Captain Abbott, his officers and men."

In a semi-official letter to Colonel Cuthbert Davidson (late Commanding the 4th Cavalry), the Resident at Hyderabad, dated 1st December, 1858, Sir Hugh Rose wrote:—

"Will you allow me to bring to your notice the claims of some gallant soldiers of your Contingent to the promotion for which I recommended them for their conduct at Jhansi? Their names are in the enclosed list.* They were mentioned specially in my despatch for storming, dismounted, a house defended desperately by a party of Wilaities and Pathans, who had escaped from the city.

"I owe so much to the assistance of your excellent Contingent, that I am very anxious that the good services of these men should not remain without the

* Some men of the 1st and 4th Cavalry.

reward, which they have so honourably earned, and I shall feel extremely obliged to you if you would have the kindness to promote them.

"I avail myself of this opportunity to thank you, my dear Sir, most cordially, for having allowed me to have command of a large portion of your force, which owes so much of its efficiency to the excellent organisation which you introduced into the corps. I shall always remember the never-failing good-will which its officers and men displayed towards myself, and the good which they did to the cause of my Queen and country."

The Resident had previously expressed his thanks and appreciation in the following General Order, dated 17th July, 1858:—

"1. The different corps and detachments of the Hyderabad Contingent, with the exception of two detachments of cavalry, having been ordered to return from the field service, on which they have been engaged in Central India, the Resident takes this opportunity of congratulating Major Orr and all the officers and men, who were employed in the above-mentioned service, on the proved devotion and gallantry with which they have upheld the name and fame of the Hyderabad Contingent in a campaign of the most arduous description.

"2. Fortunate in a General who inspired all ranks with confidence, and who never failed to lead his army to victory, the Hyderabad Contingent, as a component part of that army, has repeatedly elicited Major-General Sir Hugh Rose's public recognition of its services.

"3. Brigadier Hill and all the officers commanding corps in the Hyderabad Contingent, when mutiny and revolt broke out in Hindustan, at the unanimous desire of their men, volunteered their services wherever required by the Government of India, and nobly have they responded to the confidence which the Right Honourable the Governor-General reposed in them by his acceptance of their proferred services.

* * * * * * * *

"5. The whole Contingent were anxious to take the field, but all could not be spared from the Deccan.

"6. While waiting at Edlabad to be ordered on to meet the enemy in Central India, amid the first torrents of the monsoon, cholera broke out and devastated the camp. This of itself would have discouraged most native troops; but the reports which at that time reached the Resident gave him confidence as to the result of future events. They were all expressive of high hope, constancy and courage.

* * * * * * * *

"8. The different actions in which the services of the Hyderabad Contingent have been engaged are:—

Dhar, Piplia, Rawal, Mandesur, Rahatgarh, Madanpur, Chanderi, Garhakota, Jhansi, Betwa River, Banda, Kunch, Kalpi, Bilawa, Gwalior."

* * * * * * * *

The officers of the Hyderabad Contingent Cavalry who took part in the operations above narrated were:

Captain S. G. G. Orr, 3rd Cavalry, Hyderabad Contingent.

Captain H. D. Abbott, 1st Cavalry, Hyderabad Contingent.

Captain W. Murray, 4th Cavalry, Hyderabad Contingent.

Lieutenant H. Clerk, 3rd Cavalry, Hyderabad Contingent.

Lieutenant H. C. Dowker, 1st Cavalry, Hyderabad Contingent.

Lieutenant E. W. Dun, 4th Cavalry, Hyderabad Contingent.

Lieutenant Hastings Fraser, 4th Cavalry, Hyderabad Contingent.

Lieutenant F. Samwell, 4th Cavalry, Hyderabad Contingent.

Lieutenant H. J. E. Teed, 3rd Cavalry, Hyderabad Contingent.

Lieutenant A. A. Johnson, 1st Cavalry, Hyderabad Contingent.

Lieutenant Westmacott, 4th Cavalry, Hyderabad Contingent.

Surgeon W. Mackenzie, 3rd Cavalry, Hyderabad Contingent.

Surgeon J. H. Orr, 4th Cavalry, Hyderabad Contingent.

Assistant-Surgeon A. Sanderson, 1st Cavalry, Hyderabad Contingent.

One infantry officer of the Hyderabad Contingent Field Force, Captain Sinclair, was killed at the storming of Jhansi. The following cavalry officers were wounded or invalided:—

Captain H. D. Abbott—Wounded.
Captain W. Murray—Wounded.
Lieutenant H. Clerk—Severely wounded.
Lieutenant H. C. Dowker—Severely wounded.
Lieutenant F. Samwell—Dangerously wounded.
Lieutenant Westmacott—Severely wounded.
Captain S. G. G. Orr—Proceeded to Europe (sick).

Numerous honours and rewards were conferred on the native officers and men of the Hyderabad Contingent for distinguished conduct and conspicuous gallantry in action.

These were:—

1st Class Order of British India, with the title of Sirdar Bahadur—3.

2nd Class Order of British India, with the title of Bahadur—1, subsequently increased to 6.

3rd Class Order of Merit—19, subsequently increased to 71.

Promotions in the Cavalry.

Risaldar-majors—7.
Ressaidars—5, subsequently increased to 6.
Jemadars—18, subsequently increased to 23.
Duffadars—24, subsequently increased to 29.

In the *London Gazette* of 16th November, 1858, Major W. A. Orr, who commanded the Hyderabad Contingent Field Force, was promoted Brevet-Lieutenant-Colonel from 24th March, 1858.

Captain H. D. Abbott was promoted Brevet-Major from the same date, and Surgeon John Henry Orr was made an extra member of the Military Division of the Third Class or Companions of the Order of the Bath.

In the *London Gazette* of 22nd March, 1859, the following were given C.Bs.:—

Lieutenant-Colonel William A. Orr.
Major H. D. Abbott.
Surgeon William Mackenzie.

Lieutenant Dowker was at a later date promoted to a brevet majority on attaining the rank of captain, and was subsequently made a Companion of the Order of the Bath for his distinguished services in the campaign in Central India.

In the *London Gazette* of July 29th, 1859, Captain William Murray was promoted Brevet-Major.

Risaldar-Major Umar Khan was admitted to the 1st Class of the Order of British India, with the title of Sirdar Bahadur, as a special case, in consideration of his conspicuous acts of loyalty to the State.

Ressaidars Fazil Khan and Didar Baksh Khan were admitted to the 3rd Class of the Order of Merit.

Trumpeter Fateh Khan was promoted Trumpet-Major.

Subsequently for gallantry in an engagement with marauding Arabs and Rohillas near Bosi on the 13th November, 1859, Sowars Tankul Khan and Mahbub Beg were admitted to the 3rd Class of the Order of Merit.

A full list of rewards and their recipients is not available.

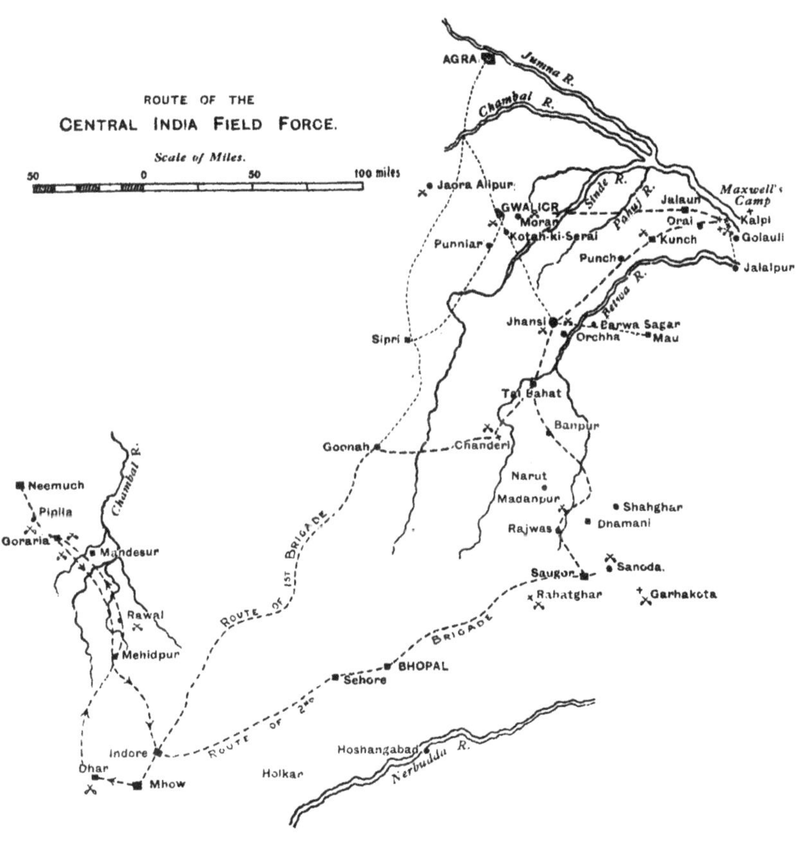
ROUTE OF THE
CENTRAL INDIA FIELD FORCE.
Scale of Miles.
50
0
50
100 miles
AGRA
Jumna R.
Chambal R.
Jaora Alipur
GWALIOR
Morar
Kotah-ki-Serai
Punniar
Sinde R.
Pahuj R.
Jalaun
Maxwell's Camp
Kalpi
Orai
Kunch
Golauli
Punch
Jalalpur
Betwa R.
Sipri
Jhansi
Barwa Sagar
Orchha
Mau
Tal Bahat
Banpur
Goonah
Chanderi
Narut
Madanpur
Shahghar
Rajwas
Dhamani
Saugor
Sanoda.
Rahatghar
Garhakota
Neemuch
Piplia
Goraria
Mandesur
Chambal R.
Rawal
Mehidpur
ROUTE OF 1ST BRIGADE
BRIGADE
BHOPAL
Sehore
ROUTE OF 2ND
Indore
Dhar
Mhow
Holkar
Hoshangabad
Nerbudda R.

It is also impossible for the same reason to give a full return of the many casualties from wounds and disease, suffered by the Regiment during this arduous campaign. There is, however, a return showing the casualties in action of the native ranks of the Hyderabad Contingent Cavalry from the battle of the Betwa to Bilayan. This is given below:

	Betwa.		Jhansi.		Kunch.		Kalpi.		Bilayan.	
	K.	W.	K.	W.	K.	W.	K.	W.	K.	W.
1st Cavalry ...	1	4 ...	1	10 ...	1	8 ...	1	1 ...	0	2
3rd Cavalry ...	2	4 ...	0	0 ...	1	2 ...	1	2 ...	not present.	
4th Cavalry ...	2	2 ...	3	8 ...	1	5 ...	4	3 ...	0	2

The 3rd Cavalry lost more heavily in the Malwa Campaign; at Dhar their casualties were two killed and five wounded, and at Mandesur one killed and nine wounded.

Taking everything into consideration, at this rate the total casualties from all causes of the Hyderabad Contingent Cavalry must have been very considerable, especially when it is remembered that the 1st and 4th Cavalry started very weak, and therefore losses with them represent a bigger percentage.

It seems practically certain that the heaviest casualties were at Rawal (see above, p. 71). In this action, 337 sabres of the 1st, 3rd, and 4th Regiments, having made a most rapid advance of 72 miles, came upon the rebels in position with guns. They had no supports within at least 60 miles, but they attacked without the slightest hesitation, completely routing the enemy and taking their guns. Their casualties however, amounted to nearly 100, including Lieutenant Samwell, 4th Cavalry.

CHAPTER IV.

THE THIRD BURMESE WAR.

THE operations of regular warfare, which took place in 1885, and led to the downfall of King Thibaw, had come to a close before the Hyderabad Contingent was called upon, in August 1886, to furnish its quota for service in Burma.

A brief résumé of the campaign from its commencement, in November, 1885, taken from Major Burton's "History of the Hyderabad Contingent," is given here, so that the situation of affairs may be understood. The subsequent operations in Burma were not of the nature of regular warfare, and did not comprise a continuous campaign. It is, therefore, difficult to give any connected account of these operations, especially as the cavalry regiments were never together, but split up into numerous detachments of the strength of a squadron, or a troop, or even a non-commissioned officer's party.

In the middle of November, 1885, the palace and redoubt of Minhla were taken, being bombarded by armed steamers from the River Irrawaddy, and attacked by the troops on land. The fall of other strongholds followed. King Thibaw surrendered before Ava was reached by the fleet, and Mandalay was occupied on the 29th November. Shwebo and Bhamo were seized before the end of the year, and by the 31st March, 1886, the country was strategically occupied by the British troops. Mandalay, the capital, was strongly held and surrounded by a cordon of posts, and there were posts at convenient distances along the Irrawaddy, from Thayetmyo to Bhamo, at Alon on the Chindwin, and from the railway terminus at Tonghoo to Myingyan and Mandalay. All Thibaw's strong forts were in the possession of the British, together with his warships, arsenals, and small-arms factory, while the greater portion of his army had been disbanded.

In February, 1886, the Viceroy of India, the Marquis of Dufferin, made a kind of triumphal progress to Mandalay, and the country appeared to have been completely conquered. The cessation of hostilities was, however, only temporary. The overthrow of the kingdom of Ava was followed by a state of anarchy throughout the country, involving much bush-fighting, which continued for some years before Upper Burma was finally pacified.

Thus, in August, 1886, when the 3rd Cavalry were ordered to Burma, the condition of affairs was officially reported as follows:—

The Burmese Army had been disbanded, but most of the men had carried off their arms, while the Civil Government had stopped. This, naturally, led to a state of anarchy, which, by the beginning of April, 1886, developed into a widespread rebellion. Several pretenders to the throne sprang up, who easily obtained a following from amongst the ex-officials, the disbanded soldiery, the disaffected generally, and the criminal classes. Every man in power with a grudge against another, took advantage of the general confusion to settle old scores, whilst the dacoit leaders, whose opportunity in Burma has ever been in troublesome times, took advantage of the unsettled state of the country, to collect their bands and plunder the districts. The military aspect of the situation thus was, that there was no organised enemy in the field, and, therefore, no particular object in concentrating large masses of troops, but the country generally was overrun by armed bands.

Experience had already proved that mere visits of flying columns to different parts of the country were insufficient, and that for its pacification and the suppression of dacoity, military posts of sufficient strength to maintain order in their immediate neighbourhood must be established; in fact, to closely occupy the whole country everywhere was an absolute necessity. This was accordingly done, and Sir George White describes the nature of the operations, from April to August, 1886, in carrying out this plan, as being:—expeditions and marches, often abortive; numerous convoys over unbridged and unmetalled roads, often through flooded paddy fields in the hot weather and rains; the maintenance of communications, telegraphic and postal, and the collection of intelligence, requiring the daily employment of constant patrols and detachments.

These duties entailed much exposure to a trying climate, and a consequent heavy sick and death-

Patrol in Burmese Jungle, 1888.

rate. The rainy season did not cause the bands of insurgents and dacoits to disperse.

The actual resistance offered to the troops, was not of much account to disciplined and well-armed soldiers; but small bodies of those soldiers had often to stand up against bands, whose numbers were estimated in thousands.

The favourite field of the dacoits was in a jungle, where the range and precision of the rifle was of little avail, and the troops often had to drive them from well selected positions, such as the walled pagodas, with which the country abounds.

The difficulties of climate and terrain, as well as those offered by an extremely elusive enemy, and the other obstacles mentioned above, were what the 3rd and 4th Cavalry had to contend with in this campaign.

In August, 1886, the 3rd Cavalry, then stationed at Bolarum, received orders to proceed to Burma; while the 4th Cavalry were moved from Mominabad to take their place.

The 3rd Infantry, Hyderabad Contingent, were also ordered on field service to Burma at this time.

Burton points out that the Hyderabad Contingent were peculiarly fitted for service of this nature: "Stationed in remote and isolated cantonments in the Deccan, they had been afforded special opportunities for practice in jungle warfare, whilst the fact that many officers of the force had been addicted to big game shooting in the forests in the vicinity of their stations, could not but add to their efficiency and to that of their men, who had been in the habit of accompanying them on such expeditions. The result proved, particularly in the case of the 3rd Cavalry, that their mode of life and training had not been without its advantages, and the skill acquired in the chase of wild beasts was found of great service in the pursuit of dacoits, through the dense and wild jungles of Upper Burma."

On arrival in Upper Burma the 3rd Cavalry, which had been posted to the 5th Brigade, commanded by Brigadier-General R. C. Stewart*, proceeded to Shwebo, where the headquarters and 318 lances were at first established, while 87 lances were sent to Shemaga.

The British officers with the Regiment were:—

Lieutenant-Colonel C. J. O. Fitzgerald, Commandant.

Captain C. E. Gubbins, Second-in-Command.

Lieutenant J. W. B. Meade.

Lieutenant E. F. H. McSwiney.

Surgeon-Major C. E. McVittie, Medical Officer.

Attached.

Captain H. M. Mason, 1st Cavalry (afterwards Commandant, 4th Cavalry).

Lieutenant E. L. Wright, 2nd Cavalry.

Lieutenant R. Wapshare, 4th Cavalry (transferred when in Burma to 3rd Cavalry).

Subsequently the Regiment was distributed in numerous detachments, at fifteen posts east of the Mu River in the Shwebo District, and ten posts west of the same river.

* Afterwards Sir Richard Stewart, Honorary Colonel 30th Lancers.

Of these posts Shwebo was the most important, and had a strong mixed garrison. Ye-U, on the right bank of the Mu River, had a large cavalry stockade, with ample accommodation for the men and horses of an entire squadron. On either side was a civil stockade with police barracks and gaol, and a large infantry and transport stockade round a number of pagodas and *phoongye* houses. Hluttaik was a small stockade, sufficiently large for a single troop. It was a most unhealthy place, and the whole detachment there were at times down with fever.

Htantabin was another small post, with accommodation for a troop.

The three last-named posts were taken over the following year by the 4th Cavalry, but Hluttaik was finally abandoned on account of its unhealthiness.

It will be understood, from this extensive distribution, that considerable responsibility was thrown on both British and native officers, especially in view of the paucity of the former.

The operations carried out so successfully by the 3rd Cavalry, besides those of a desultory nature, may be divided into distinct phases, including (1) the pacification of the Ye-U district, (2) the pursuit of the dacoit leader, Hla-U, (3) the occupation of Wuntho, (4) pursuit of the Hlagaing princes in the Hnaw Forest.

The first of these operations was carried out by a squadron under Colonel Fitzgerald, in co-operation with another squadron, under Captain Gubbins, from the end of October to November, 1886.

In November, 1886, operations were undertaken, against the chief Hla-U, by four columns between the Rivers Mu and Chindwin.

The Ye-U column, under Colonel Middleton, consisted of—

3rd Cavalry	115 lances.
King's Own Light Infantry ...	50 rifles.
Mule Battery	2 guns.
21st Madras Infantry	100 rifles.

The 3rd Cavalry detachment was the squadron under Lieutenant Wapshare, which had been posted at Ye-U. Colonel Fitzgerald also accompanied it.

Hla-U fled on the approach of the pursuing columns, and was followed by the 3rd Cavalry. For nine days they pressed close on the fugitive, traversing dense jungle at the rate of twenty miles a day, bivouacking by night and tracking by day. The operations were not without result, although the leader was not captured, for between the 20th and 28th November his followers dwindled from some 330 men to 30.

At the same time the 2nd Squadron, under Lieutenant Wright, and the 3rd Squadron, under Lieutenant McSwiney, were co-operating by patrolling both banks of the Mu River and doing some very useful work. Lieutenant McSwiney had several successful skirmishes, killing a number of dacoits and their leaders and capturing many more, also a quantity of arms and some cattle. In December Lieutenant Wapshare, after a rapid march of 56 miles, made an important capture of arms and cattle. In January Colonel Fitzgerald again resumed operations against Hla-U, but, in spite of the strenuous exertions of his pursuers, the dacoit chief again eluded all attempts to capture him. However, nearly 500 of his followers, 550 guns, and upwards of 200 dahs and spears were either captured by, or surrendered to, the troops of the 5th Brigade before the end of January, 1887.

Hla-U was eventually killed by one of his own followers.

The district of Wuntho is situated to the west of the Bhamo district, north of the Shwebo district, and east of the Upper Chindwin. The Tsawba having refused to acknowledge British supremacy in Upper Burma, it was decided, in December, 1886, to send a force to seize and temporarily to control his capital. An advance was made in January, and at the same time a second column, including two squadrons of the 3rd Cavalry, under Captain Gubbins and Lieutenant McSwiney respectively, was sent to co-operate from the south.

There were several successful encounters with the enemy, in which they were routed with loss; Wuntho was entered and the Tsawba fled. Captain Gubbins with 104 lances of the 3rd Cavalry, in consequence of a report received of an attack by the enemy, marched towards Hluttaik and encamped at Nyaungu on 2nd February. At 2 a.m. his camp was attacked by two or three hundred dacoits; the two sentries, at opposite ends of camp, were shot down simultaneously; one was killed, but the other, Bazid Khan, although severely wounded, remained at his post and continued to return the enemy's fire. For this brave action he was afterwards admitted to the 3rd Class of the Order of Merit. The attack lasted some time; a duffadar was also wounded and a horse and three ponies shot. Two days later Captain Gubbins surprised these same dacoits, killing sixteen of them.

Towards the end of January Lieutenant Wapshare, with 25 lances, 3rd Cavalry, and 50 rifles, 21st Madras Infantry, proceeded to Nabekgyi, which had been attacked by dacoits under Prince Thinka-yaza and Boh Hantha; pushing on thence with 17 lances, he surprised the Prince's camp at Chaungzone, killed three dacoits, and took 17 prisoners. On February 4th, he had another sharp skirmish with dacoits in a dense bamboo jungle, killing eight. The 3rd Cavalry had two men and two horses wounded in this action.

On 24th February Lieutenant Wapshare, with 40 lances, accompanied Mr. Porter, the Civil Officer, and after capturing a Boh and two branded dacoits at Ganzama, pursued Boh Hantha into the Hnaw Forest, and fought an action with the whole gang of 250 dacoits at Puluswa. Boh Hantha and 39 dacoits were killed and many wounded. Lieutenant Wapshare received the thanks of the Chief Commissioner for this service.

During the months from March to June, there were a number of smaller engagements, in one of which Jemadar Sirdar Khan was mortally wounded; he lived long enough to receive his promotion to Ressaidar, and a silver cup subscribed for by the detachment of the King's Own Light Infantry at Hluttaik, in recognition of his gallant services.

In June Lieutenant McSwiney, with 31 men,

fought a successful action near Theo, killing two leaders and nine other dacoits. For this he received the thanks of the General Officer Commanding the Upper Burma Field Force, and Jemadar Abdul Aziz Khan and Trooper Ghafur Khan were both admitted to the Order of Merit of the 3rd Class for conspicuous gallantry.

On July 14th Colonel Fitzgerald received orders to proceed against the Hlagaing Princes, who had collected a large following in the Kani district, and were making raids into Ye-U. He at once decided to advance in four columns of 20 to 30 lances each, to try and surround the Princes, and prevent their crossing the Chindwin.

Early on the morning of the 15th the columns were in motion through the dense forest, where communications were maintained with difficulty.

On the night of the 18th June, Colonel Fitzgerald made a forced march from Imbaung to Yin, and, at 8 a.m. next day, came up with the younger Prince's forces, marching in a southerly direction near the village of Kantha. The enemy, over 200 strong, took up a strong position. Colonel Fitzgerald, without hesitation, charged and routed them with a loss of nine killed, many wounded, and some prisoners. The Cavalry had had a very hard day, having marched 60 miles in 24 hours, and been 18 hours in the saddle. For this action Colonel Fitzgerald received the thanks of the Major-General Commanding the Upper Burma Field Force.

The 3rd Cavalry returned to India in January and

February, 1888, on relief by the 4th Cavalry. The Chief Commissioner, Sir Charles Crosthwaite, in the course of an address at Ye-U, publicly acknowledged their services.

Their strength during the campaign was 497 of all ranks, 367 followers, 529 horses, and 295 ponies. Two men were killed in action, one native officer died of his wounds, 12 men were wounded, 20 died of disease, and 26 were invalided.

Casualties among horses.—Two hundred and twenty-one died of "Kumri," mostly in the months of February, March, July, August and September, 38 from other diseases, four were killed in action, and seven wounded. Two died on the voyage.

The 4th Cavalry left Bolarum for Upper Burma organised temporarily in four squadrons. Each squadron moved by train to Madras, thence by Indian Marine Transport to Rangoon, and up the Irrawaddy by river steamer, as follows:—

1st Squadron, with Lieutenant-Colonel Walker, Commandant, Surgeon Kellie, Medical Officer, and Lieutenant Fitzroy Johnstone, Adjutant, left 9th January, 1888, and arrived at Ye-U 10th February.

2nd Squadron, with Captain F. R. B. Knox and Lieutenant T. D. Leslie, 2nd Cavalry, left 10th January, and arrived at Mandalay 14th February.

3rd Squadron, under Lieutenant Wyllie, 1st Cavalry, arrived at Mandalay 18th February.

4th Squadron, with Lieutenant-Colonel Cummins, left 29th January, and arrived at Mandalay 27th February. Lieutenant S. M. Mason and Lieutenant H. R. Beddoes accompanied it.

Facing Page 155.

A Troop at Ye-U, 1888.

It should here be mentioned that the Regiment relieved the 7th Bengal Cavalry in Upper Burma, as well as the 3rd Cavalry Hyderabad Contingent.

After arrival the distribution during 1888 at first was:—

1st Squadron at Ye-U.

2nd and 3rd Squadrons at Sonewa and Myotha in the Ava district.

4th Squadron—one troop at Maymyo in the Shan Hills under Jemadar Karim Yar Khan, and one troop in the Sagaing district, under Lieutenants S. M. Mason and Beddoes.

The districts had now to some extent been reduced to order, but much still remained to be done. The dacoits had gained experience in the art of war; no longer did they rashly venture out in the open in the presence of cavalry; on the contrary, they chose thick jungle and other ground unfavourable to that arm, as the scene of their attacks. It must be remembered that formerly there were no horses in Burma, for only the small Burma pony existed in that country, and so a cavalry regiment was an entire novelty to the inhabitants. Their favourite tactics were to lay an ambuscade in the thickest jungle they could find, and when a column approached suddenly pour a volley into it, and then retreat with the utmost rapidity. Under these circumstances, the best tactics for the cavalry were to instantly gallop at the spot from whence the shots were fired, and in this way it was often possible to account for some of the dacoits in the act of making off.

Another plan they evolved, was to establish an

elaborate system of "bush telegraph," by means of men up trees. By this the movements of cavalry and other columns were signalled for miles ahead. This necessitated the troops moving a great deal by night; for example, if intelligence was brought that a band of dacoits had assembled in a certain village a force would start, march throughout the night, surround the village, and rush it at daybreak. Of course, in addition, every precaution was taken to keep the movements secret; the cavalry never used pennons on their lances; in fact, the points were usually dulled, to prevent them flashing. It may also be added here, that a stop, about six or eight inches down from the lance point, was also found very useful, as otherwise the lance used to go right through a man and the shaft get broken. To sum up, the enemy, though now not nearly so numerous, had become much more skilful and elusive.

From the foregoing the class of operations which were performed by the 4th Cavalry may be understood.

The 1st Squadron were at Ye-U, under Lieutenant-Colonel Cummins. A detachment was sent to Hluttaik, where it remained some time, but the place was found so unhealthy that this detachment was eventually removed to Htantabin. The whole squadron was actively employed in the Ye-U district with Lieutenant-Colonel Cummins and Lieutenant Johnstone. Numerous expeditions were carried out, some going as far north as Wuntho; a number of dacoits were killed and captured, and the country was gradually brought under control.

At the same time the 2nd and 3rd Squadrons were

operating from Sonewa and Myotha, in the Kyouksai and Ava districts, under the command of Colonel Walker and Captain Knox respectively; with them were also Lieutenants Wyllie and T. D. Leslie. On 2nd March, when encamped at the village of Deweyla, Colonel Walker, who had with him part of the Simun column, consisting of 54 lances of the 3rd Squadron and a detachment of the 15th Madras Infantry, received information that the notorious leader Boh Tok, with a large body of followers, was in some thick jungle near the village of Thepan. He accordingly marched at 5.30 next morning, but found that the enemy had retired to a position at Iswa, four miles south of Thepan.

On approaching the enemy's stronghold the force was divided into two parties, one of which, under Lieutenant Wyllie, was sent by the right path leading to the dacoit position, while Colonel Walker led the other by the left path. An advanced picquet of the enemy was first surprised; a farther advance was then made, when suddenly from a nullah the enemy opened a hot fire on Colonel Walker's party. Ressaidar Mir Muzaffar Ali was shot dead, and his orderly, Nasib Ali, severely wounded; this trooper behaved with great gallantry, remaining by the ressaidar's body, even after he was wounded, and attempting to drag it under cover; he richly deserved the Order of Merit. The ground in front was found to be unrideable; the cavalry then dismounted with swords and carbines, and rushed the position on foot, supported by the infantry, who had now come up. The enemy gave way, and owing to the denseness of the jungle a number of them escaped.

At 5 a.m. on March 20th Lieutenant Wyllie, with 30 lances, left the village of Nyedo in the Kyouksai district, and, about a mile from the village of Mojudwin, surprised a party of dacoits, who opened fire. The cavalry instantly charged and routed them, killing eleven of them, including Boh Pyau, and capturing two badly wounded. Risaldar-major Muhammed Muzaffar Khan was slightly wounded in this affair.

Lieutenant Wyllie had another skirmish with dacoits in April, and many other encounters of a similar nature took place.

The 2nd Squadron, under Captain Knox, was engaged in the country to the west of the Moozatoung Range in the Ava district, and did a great deal of arduous and good service. When in April Colonel Walker, in command of the Simun Flying Column, resumed the pursuit of Boh Tok, a portion of the 2nd Squadron co-operated. Boh Tok was eventually driven from his position in the Moozatoung Range; many of his followers were killed and captured, while he himself, flying thence, fell in with and was killed by a detachment of the Rifle Brigade.

A party of 13 lances of the 4th Cavalry, under Duffadar Muhammed Fazl Khan, on 25th July, took part in an engagement with the noted leader, Boh Shwe Yan and about 60 of his gang, near the village of Thagyin in the Ava district. Lieutenant Manogue, of the 2nd Battalion Royal Munster Fusiliers, was in command of the whole force, which included twenty-five men of the Munster Fusiliers, in addition to the cavalry detachment. The enemy fought well for about half an hour, but were finally routed, Boh Shwe Yan and ten of his men being killed and a number wounded. Lieutenant Manogue, in his despatch of 26th July, remarks:—

"Both the Sowars and Munster Fusiliers worked exceedingly well together; every man appeared quite cool, and every shot seemed fired with a deliberate aim."

Duffadar Muhammed Fazl Khan was wounded in the right leg in this skirmish.

Captain Knox, who planned the expedition, Lieutenant Manogue, and the party under his command received the thanks of the Chief Commissioner and of the Major-General Commanding the Upper Burma Force. The reward of Rs.1,000, set on the head of Boh Shwe Yan, was distributed among the men of the Munster Fusiliers and 4th Cavalry, who took part in the affair.

On 23rd May the 3rd Squadron reached Mandalay, and the Headquarters of the Regiment, with Colonel Walker and Surgeon Kellie, were established there.

The 2nd Squadron remained at Myotha and Maghee, sending detachments across the Irrawaddy, to co-operate with the detachments of the 4th (extra) Squadron and other troops employed in clearing the Sagaing district of the numerous bands of dacoits that infested it. A great deal of good work was done here, and the excellent results achieved by the detachments under Lieutenants S. M. Mason, T. D. Leslie, and Beddoes were frequently brought to notice by Colonel Symons, under whose orders they were operating.

Lieutenant Mason was specially thanked for his services in reducing this district to order.

At the same time Captain Knox and Lieutenants Johnstone and Wyllie were similarly employed about Myotha and Maghee, on the other bank of the river.

It may here be mentioned, that the other cavalry regiment with the Upper Burma Force at this time, was the 1st Madras Lancers. They were working the country farther south.

Jemadar Soobhan Khan, who was at Htantabin with a detachment of the 1st Squadron, shortly after midnight on 8th October, 1888, received intelligence that a strong band of dacoits were attacking the village of Nagaram. He turned out at once with 25 lances, and after a rapid night march reached the village at daybreak, only to find that the dacoits had set it on fire and retired, carrying with them a number of prisoners and cattle. The pursuit was immediately taken up, and after 25 miles the dacoits were overtaken. It was then 11 a.m., and a running fight was kept up till 3 p.m. The enemy, who numbered from sixty to eighty men, were totally routed, and lost fourteen killed, including two identified and two suspected Bohs.

Trooper Imam Ali Khan and another man were wounded, also three horses. Twenty-five head of cattle, as well as the men and women taken prisoners by the dacoits, were recovered. This affair reflected great credit on Soobhan Khan and his men, as it took place in dense bamboo jungle, quite unsuited for cavalry operations. Brigadier-General Wolseley, C.B., A.D.C., characterised the jemadar's conduct as "prompt and gallant."

On 2nd January, 1889, Lieutenant Beddoes, with a jemadar and nine lances, and a detachment of mounted police, surprised the camp of the dacoit, Boh Ngou, near Gumzeit, and killed him and two of his gang. Trooper Sanoo Khan, a gallant old soldier, was shot dead while in the act of spearing Ngou; the Boh, with the lance actually through him, succeeded in shooting his assailant. Two troopers were also wounded, and one, Muhammed Khan, died of his wounds the same day. In his report of this skirmish to the General Officer Commanding 1st Brigade, Colonel Symons recommended to his favourable notice the arrangements of the officer in command of the party and the admirable conduct of the men. The large reward, which had been offered by Government for the capture of this celebrated Boh, was divided up among the detachments.

At the end of 1888 the Regiment was distributed as follows:—

The 3rd Squadron, with headquarters, was at Mandalay, and with it were Colonel Walker, Lieutenant F. Johnstone, Lieutenant T. D. Leslie, and Surgeon Kellie. Quarters were obtained in some **phoongye kyoungs*, at the back of Mandalay Hill, at some distance outside the palace moat. The 1st Squadron had one troop at Htantabin, with Lieutenant-Colonel Cummins, and one troop in the

* Houses of the Buddhist priests about the pagodas.

cavalry stockade at Ye-U, under Lieutenant Stotherd.

The 2nd Squadron were mostly at Myotha and Maghee, under Lieutenant Wyllie. Captain Knox, who had been in command here, left for India towards the end of November, to take up the appointment of Second-in-Command of the 3rd Cavalry. Some detachments of the 2nd Squadron were in the Sagaing District under Lieutenant S. Mason, who had also with him Lieutenant Beddoes and detachments from the extra 4th Squadron, the remainder of which was still at Maymyo, in the Shan Hills, under Jemadar Karim Yar Khan.

In February, 1889, the Regiment left for India in three squadrons. Each squadron moved separately at intervals of from two to three weeks. The first to go was the 3rd Squadron, with Colonel Walker and headquarters; next the 1st Squadron, with Lieutenant-Colonel Cummins; finally, the 2nd Squadron, with Lieutenant S. Mason. In each case the procedure was the same; the men were embarked on one of the Irrawaddy River steamers, with the horses and ponies on board flats lashed alongside.

The steamers only travelled by day, and tied up to the bank at night, when, if possible, the horses and ponies were landed and picketted, as they suffered a good deal from mosquitoes, and used to stamp and kick the whole night along. At Rangoon the squadron was transhipped to the Indian Marine Ship "Canning" and taken to Madras. In those days landing at Madras was difficult; each horse had to be put in slings and hoisted by a derrick over the ship's side into a surf-boat, which was rowed as near to the beach as possible, and the horses were then turned out into the surf and scrambled on shore, where the men were waiting to catch them. The Squadron then encamped on the glacis of Fort George. The railway journey from Madras was a very slow business; the troop trains only travelled at night; in this way the first night's journey only went as far as Gooty, where the day was spent in the rest camp; the second was to Wadi Junction, and the third to Sholapore, whence there was a hundred miles' march to Mominabad, to which station the Regiment now returned.

When the Regiment left Upper Burma a complimentary order was issued by Brigadier-General Wolseley, C.B., A.D.C., in which he expressed his high opinion of it, and the way it had performed its duties.

The following British officers served with the 4th Cavalry during the campaign:—

Lieutenant-Colonel J. G. D. Walker, Commandant.

Lieutenant-Colonel J. T. Cummins, Second-in-Command.

Captain F. R. B. Knox.

Lieutenant R. F. M. Johnstone, Adjutant.

Lieutenant S. M. Mason.

Surgeon G. Kellie, Medical Officer.

Probationers.

Lieutenant H. R. Beddoes, 7th Queen's Own Hussars.

Lieutenant E. A. W. Stotherd.

Attached.

Lieutenant F. Wyllie, 1st Cavalry H.C.

Lieutenant T. D. Leslie, 2nd Cavalry H.C.

There were thirteen native officers present, 449 rank and file, and 462 horses.

The casualties were:—

Killed in action, one native officer and two men.

Wounded, one duffadar and seven men.

Died of disease, nine men.

Invalided to pension, three men.

Invalided to India, 45 men.

Horses—

Died or were destroyed from	*Kumri* ...	148
Ditto	*Surrah* ...	40
Other causes	... ...	8
Total ...	... ...	196

Kumri caused great devastation among the horses of every cavalry regiment that went to Burma. Very often horses affected by it would appear, when in the stable, to be perfectly well, but when taken out paralysis of the hind-quarters would be at once apparent. Most of the disease seemed to originate in the cold damp of the rainy season, but in spite of the numberless cases, neither the cause nor the cure were ever accurately determined. The usual treatment was a blister on the spine about the croup, but it was practically never successful. Government decided that compensation was allowable under the same regulations as for an officer's charger; the consequence was that for every horse, which died or had to be destroyed from *Kumri* or *Surrah*, a compensation committee had to be held. The mortality among the baggage ponies from these causes was also very great. Burmese ponies were seldom affected by *Kumri*. Additional casualties were a number of horses wounded in action. In most cases the wounded animals quite recovered. In one instance two horses, ridden by men forming an advance party to a column, were both wounded by a volley fired from an ambuscade. One got a bullet in the withers, which was successfully extracted, but the other had a bullet lodged in the head, which could not be got out; however, it was taken back to India, and did several years' more useful work in the Regiment, apparently suffering no inconvenience from the bullet.

During the Burmese War the Regiment was mounted almost entirely on Arabs.

Burton, in his summary of the operations, says:—

" Although the losses in this campaign were slight, the duties so well performed by the troops were tedious and harassing, as will be understood from the description of the nature of those duties in the early part of this chapter. To the pacification of Upper Burma, carried out during the years 1886-1889, the Regiments of the Hyderabad Contingent largely contributed." . . . . . .

A curious feature of the operations was that tents, though of course taken to Burma, were seldom, if ever, used. Officers and men lived in the *phoongye kyoungs* and Burmese houses; in any case, when on columns and patrols against such an extraordinarily mobile enemy, tents were out of the question.

The following rewards were given for this campaign:—

Brigadier-General R. C. Stewart and Colonel C. J. O. Fitzgerald were made Companions of the Bath.

Lieutenant-Colonel J. T. Cummins and Lieutenant E. F. H. McSwiney were made Companions of the Distinguished Service Order.

Captain C. E. Gubbins was promoted Brevet-Major; Risaldar-Major Muzaffar Khan, 4th Cavalry Hyderabad Contingent, was made Companion of the Order of the Indian Empire; and Risaldar-Major Abdul Karim Khan, 3rd Cavalry Hyderabad Contingent, was admitted to the Order of British India.

CHAPTER V.

UNIFORM.

The uniform of the 4th Cavalry was, and always has been, the same as the other three regiments of the Hyderabad Contingent Cavalry. The following is the description of the dress laid down for the European officers of the Nizam's Cavalry in May, 1827.

Full Dress.

Jacket.—Dark green, rounded shell to fit close, and fasten down the front with hooks and eyes, the cuffs and collar white, the collar straight in front, three inches deep, and fastened with hooks and eyes; the body to be edged with white binding. Three rows of regimental buttons. The jacket to be trimmed with flat gold lace or braid.

Trousers.—White linen or calico, made loose, cut below to fit the boot, with strap underneath it.

Overalls.—Dark green, with a row of gold lace down the outer seam two inches broad; cut to fit the boot, with strap underneath it.

Boots.—Wellington.

Spurs.—Steel with necks two inches long, screwed to the boot.

Helmet.

Girdle.—Crimson and gold, three inches broad.

Sabre.—Madras Cavalry regulation, with half-basket hilt and steel scabbard.

Sword knot.—Crimson and gold.

Cravat.—Black silk.

Gloves.—White leather.

Pouch.—Metal gilt, with ornamental wreath round the border, as in Madras Cavalry, a silver star in the centre, with "His Highness the Nizam's Cavalry" in a scroll in the centre.

Pouch belt.—Plain gold, two and a half inches broad; gilt plate in front, with chain and prickers attached to a gilt button.

Waist belt and slings for sabretache.—Plain gold lace, the former one and a half and the latter three-quarters of an inch broad, fastened in front with plain gold hooks, plain gilt buckles, slides and swivels to the slings.

Sabretache.—Dark green broad cloth, with a border of plain gold lace two and a half inches broad round it, a gilt star in the centre, with "His Highness the Nizam's Cavalry" in a scroll in the centre.

Undress.

Jacket.—Dark green, facings of the same colour, trimmed with black silk lace, and buttons.

Trousers.—Dark green, made loose, and cut below to fit the boot, with strap underneath; a row of black silk lace on the outer seam two inches broad.

Forage cap.—Dark blue cloth, with gold band two inches broad, peak in front, and oilskin cover.

Facing Page 168.

Hyderabad Contingent Cavalry.
British Officer.
1845.

Boots.—Wellington.

Spurs.—Steel as in full dress.

Sabre.—In black leather scabbard, lacquered iron shoe at the end.

Sword Knot.—Black leather.

Waist-belt.—Black patent leather, one and a half inches broad, plain gilt ornaments, with frog and strap for the sword.

Pouch and belt.—Black patent leather belt, three inches broad, plain gilt plate, chain and prickers, and buckle and slide behind. Pouch to contain twelve rounds, with gilt star on the back of it. A waist-belt attached to the pouch, three-quarters of an inch broad, with gilt hook in front.

Cravat.—Black silk.

Gloves.—White leather.

Cloak.—Blue, lined with scarlet.

Saddle.—Plain Gibson's Cavalry regulation saddle and bridle, with dark green cloth valise.

After the force became the Hyderabad Contingent the full dress for British officers was changed. It then became a green Hussar tunic with gold braid, very similar in pattern to the tunic worn by Hussar regiments at the present day. The breeches were of white melton cloth, worn with black knee boots, cut with a V in front.

The belts were of gold lace, with a red stripe down the centre, and mounted on red morocco leather.

The chains and prickers were gilt, as was also the pouch, which had on it the monogram "H.C.C.," in silver, surmounted by a crown. The sabretache was of green cloth, edged with broad gold lace, and bearing the monogram "H.C.C." in gold. It was mounted on red leather.

White doe-skin gloves were worn.

The helmet was white, with the edges bound in gilt metal, and a blue and gold embroidered *loongi* end worn round it.

In full dress mounted, a leopard skin, edged with scarlet, was worn over the saddle. A throat plume, red in the centre and white outside, was used by the 4th Cavalry on all parades. The throat plumes worn by the 3rd Cavalry at this time were all red, and those of the 1st and 2nd Cavalry all white.

The undress uniform was similar to that of the British Cavalry at that period, except that it was dark green, viz., a patrol jacket, covered with mohair braid, and falling lapels, and overalls with double gold lace stripes. The forage cap was the round cavalry pattern, with broad gold band and no peak.

The mess dress was of Hussar pattern, and had a red waistcoat, with bars of gold lace. The Regimental *cummerbund* at this time was red. The 3rd Cavalry, after the Burmese War, wore a yellow one.

The present Lancer tunic of rifle green was introduced after 1890, when the designation of the regiments was changed. White facings were again adopted, and the whole uniform, including mess dress, became the usual Lancer type.

British officers first wore shoulder chains on *khaki* uniform in 1889. At one time they wore a brass-bound *khaki* helmet, with a blue and gold *loongi* end round it; this was replaced by a *khaki* helmet, without the brass binding, and covered with a plain unembroidered *loongi* end. Finally, the perfectly plain white and *khaki* helmets, as now worn, were adopted.

Up to 1877 the men wore the *ulkaluk* and girdle, and the *mundêêl;* then their uniform was changed to a blue serge blouse, with red *cummerbund*, and a blue *loongi* and *kuzzlebash*. The blue blouse was soon changed to one of rifle green, which was the immediate predecessor of the dark green *kurta* now in use.

Sam Browne belts were issued to all native ranks in June, 1884.

The field service dress, worn by the Regiment during the Burmese War, was a *khaki* blouse, red *cummerbund*, *barbal* dyed breeches, putties, and blue *loongis*, with leather *kuzzlebashes*.

This dress was then always worn at camps of exercise and field days, but, in 1901, an entirely *khaki* kit, including *loongi* and *kuzzlebash*, was adopted. With this Stohwasser gaiters were worn, but these were later replaced by *khaki* putties.

It may not be out of place to add here that formerly each squadron had a standard. These were of dark green silk, fringed with gold. In the centre was a wreath of laurels, in gold embroidery, enclosing the inscription in *urdu* character, " Allah-i-Akbar " (God is great); there was also the numeral 4, in English and *urdu*, in gold.

" Allah-i-Akbar " is the commencement of the Mohammedan battle cry before attacking.

The poles to which the standards were attached had broad spear heads and ornamental carving.

Two standards and poles are preserved in the regimental mess.

There was also what was called an undress standard for each squadron. These were latterly always kept in the regimental store, but disappeared when the Regiment was moving from Mominabad to Bolarum, in 1889.

Facing Page 172.

British Officer, Drill Order.

CHAPTER VI.

SIR JOHN GORDON OF PARK AND THE GORDON REGIMENTS.

THE Gordons of Park were descendants of Jock Gordon of Scurdargue, and cadets of the line which settled in Cairnborrow.

From the Gordons of Cairnborrow, in the seventh generation, a Sir John Gordon became the first baronet of Park; he died in 1713. Sir William Gordon of Park, the third baronet, when the Jacobite Rebellion broke out in 1745, joined the Pretender's standard, serving as Lieutenant-Colonel of Lord Ogilvie's Regiment. He fought at Culloden, and after a narrow escape of capture, managed to reach the Continent, where he took post under the Emperor of Germany, who allowed him and his heirs "the rank of the first-class nobility of Hungary." Sir William was attainted as a Jacobite, but the attainder was afterwards reversed.

To save the estate, however, Sir William made it over to his half-brother, Captain John Gordon, of the Marines, who continued to keep possession of it.

His son, Sir John James Gordon, fourth baronet, after a varied career in the 9th Foot, in France, and finally in the service of the H.E.I.C. in India, was killed in action at Bassein on 10th December, 1780, where he was said to have shown considerable gallantry.

Sir John Bury Gordon, fifth and last baronet of Park, was a posthumous child, born in Banff on April 5th, 1781, after his father's death at Bassein. It appears that his father sent his wife home to his relations in Banff, when the campaign against the Mahrattas was threatening.

Sir John's early career started well; at the age of 14, probably through the influence of Lord Fife, he was given an Ensigncy without purchase in the Coldstream Guards. From this point onwards his career lay wholly in soldiering. He spent the last 23 years of his life in India, and, during that period varied his services with the 13th Light Dragoons by taking a post under the Nizam of Hyderabad.

Sir John first appears with the Hyderabad Contingent in February, 1822, when Fort Mohun was surrendered to a force of the Nizam's Army under his command. In the same year he was appointed to the command of the Ellichpur Horse (afterwards the 5th Nizam's Cavalry), and in 1826 he raised, at Mominabad, the 4th Regiment of the Nizam's Cavalry, and became the first commander of the corps. This Regiment is the present 30th Lancers (Gordon's Horse).

It was not until 1804, when his distant cousin, John Gordon of Park died, that Sir John Bury Gordon took up the baronetcy, for the Cobairdy line not only had possession of the ancestral estates, but assumed wrongfully the baronetcy.

Sir John died without issue at Madras, 23rd July, 1835. A monument there calls him "Major, 13th Light Dragoons, commanded the Bolarum Division of the Hyderabad Contingent." On his death the baronetcy became extinct.

Mr. J. M. Bulloch, to whom the Regiment is indebted for the above particulars, and also for the following account of the Gordon Regiments, writes: —"It is enough to say, that the Gordons of Park had a full share of the dash of the devil, which has made the Gordons such excellent soldiers and specially connected with cavalry. . . .

"Concentrating our attention upon the family of Gordon alone, we find that, during the past century and a half, seven different corps have been raised by officers bearing that name.

"(1) 1759-65.—'89th Regiment'; raised by the fourth Duke of Gordon. This Regiment went to India and took part in the siege of Pondicherry. It was disbanded in 1765.

"(2) 1778-83.—'Gordon, or Northern Fencibles'; raised by the same, and disbanded in 1783.

"(3) 1793-96.—'Northern Fencibles'; raised by the same, and reduced in 1796. The national occasion which called into being this Regiment, was the declaration of war by France on 1st February, 1793.

"(4) 1794.—The '100th,' afterwards the '92nd,' and now the 'Gordon Highlanders.'

"Although these four regiments were, strictly speaking, territorial efforts in soldiering, the military spirit fostered throughout the wide dominions of the ducal family, resulted in the creation of a large number of officers, those of the name of Gordon alone standing at nearly 1,400.

"It must, therefore, be understood, that there was a very large substratum of the purely family affair in the four regiments, which the Duke of Gordon raised.

"By far the most important regiment raised by the Duke was the Gordon Highlanders.

"(5) 1826.—'30th Lancers (Gordon's Horse),' in the Indian Army; raised by Sir John Bury Gordon, fifth and last baronet of Park, who carried on the regiment-raising traditions of his house by creating a corps of horse. Thus, although Sir John left no issue, his name is perpetuated by this Regiment.

"(6) 1846.—'15th Ludhiana Sikhs,' Indian Army; raised by Major Patrick Gordon, of the Cairnfield family. This officer was a distant cousin of Sir John Bury Gordon, being descended from the daughter of Sir John Gordon, first baronet of Park.

"(7) 1857.—'Gordon's Volunteers'; raised by Lieutenant John Gordon, of the 19th Bombay Native Infantry. This was a temporary corps, which did good service in the Indian Mutiny. It was composed of natives of India."

APPENDIX.

OFFICERS WHO HAVE SERVED IN THE REGIMENT.

(A) Officers who have Commanded the Regiment.

(B) Officers who have Served in the Regiment (in Chronological Order).

(C) Officers Serving in the Regiment on January 1st, 1911.

(A) OFFICERS WHO HAVE COMMANDED THE REGIMENT.

Gordon, Sir John, Bart.

(*From the History of the XIII. Hussars*). Ensign, Coldstream Guards, 1st April, 1796; Lieutenant and Captain, do., 20th September, 1799; retired, 30th November, 1806; Cornet, 22nd Light Dragoons, 24th January, 1812; Lieutenant, do., 5th November, 1812; Captain, 13th Light Dragoons, 13th June, 1820; Major, do., 18th July, 1834; died at Madras, 23rd July, 1835. Carried the King's Colour at the landing at the Helder under Sir Ralph Abercrombie, 27th August, 1799; present at battle of Krabend, 10th September, 1799, and again carried the King's Colour. Similarly at Schagenburg, 19th September, 1799. Present at Burgen, 2nd October, 1799. Landing in Egypt under Abercrombie, commanded a company of the Coldstream Guards, March 8th, 1801; present at the battle on the same day, and at the Battle of Alexandria, 20th March, lastly at the battle westward of Alexandria, August 22nd, 1801. Wounded by the bursting of a shell (left shoulder) at Alexandria (medal for the campaign in Egypt in 1801).

In 1826 he raised the 4th Cavalry, and became its first Commandant.

Byam, Adolphus Elizabeth.

Born September 5th, 1805. Educated at Addiscombe, 1819-21. 2nd Lieutenant, June 9th, 1821. Lieutenant, June 10th, 1821. Captain, September 1st, 1831. From 1823-26 he was Lieutenant in the Corps of Artillery at St. Thomas Mount. First appointed to the Nizam's Army on February 1st, 1826. From 1827-30 he was extra assistant to the Resident at Hyderabad. He commanded the 4th Cavalry as Captain in the Nizam's Army from 1831-39.

In 1836 a party of the 4th Cavalry, under Captain Byam, marched from Secunderabad to Berhampore, a distance of 588 miles, in 31 days, arriving on the 4th November. Between November 15th and 18th he surprised three rebel chiefs and killed two of them and several of their followers, and wounded or captured others. For the above services Captain Byam and the 4th Cavalry were commended by the Governor and Council of Madras.

In 1839 he was Military Secretary to the Resident of Hyderabad, with the rank of Major; his rank in the East India Company's Army was Captain, Madras Artillery.

He died at the Cape, November 10th, 1839, aged 34, when on leave.

Garstin, Robert.

Born, August 14th, 1799. Son of Captain Garstin, 65th Regiment of Foot. Educated at the Military College, Woolwich. Cornet, February 13th, 1821, 2nd Madras Cavalry. Lieutenant, May 1st, 1824, and Captain, 2nd Madras Cavalry, 1836. Services placed at the disposal of the Resident of Hyderabad, September 22nd, 1826. Attached to the 3rd Regiment of Cavalry, Nizam's Army. Promoted to the command of the 4th Cavalry, Nizam's Army, March 13th, 1839. On sick leave to New South Wales and the Cape for two years from June, 1841. Resigned the Nizam's service, July 22nd, 1844. Appointed Government Agent at Chepauk and Paymaster of Carnatic Stipends, August 17th, 1847. A.D.C. to the Governor of Madras, and subsequently to the Commander-in-Chief of Madras in 1848. Officiated as Military Secretary to the Commission of Mysore, 1849. Promoted Major, September 4th, 1851. Proceeded to Europe on sick leave, June 4th, 1852.

Davidson, Cuthbert.

Born, May 24th, 1810. Cadet, May 11th, 1828. Ensign, April 22nd, 1829. Lieutenant, January 4th, 1832. Captain, September 8th, 1842. Major, May 10th, 1852. Lieutenant-Colonel, May 31st, 1857.

First appointed to the Nizam's Army, 1st January, 1836. Adjutant, 3rd Cavalry at Mominabad, 1838-40. 1840-41, attached to 1st Cavalry. 1841-44, commanded 4th Cavalry. 1844, with 1st Cavalry. 1845-48, commanded 3rd Cavalry at Goolburgah. 1848-50, he was Acting-Assistant to the Resident at Hyderabad.

Appointed Resident at Baroda, November 29th, 1855. Appointed Resident at the Court of H.H. the Nizam, March 13th, 1857.

Strange, William Robert.

Born, February 7th, 1809. Cornet, January 8th, 1826. Lieutenant, July 30th, 1827. Captain, March 22nd, 1842. Major, May 23rd, 1854.

First appointed to the Nizam's Army on January 31st, 1828. 1838, Brigade-Major of the Cavalry Division. In 1841-42 he was "in charge of the 5th Cavalry," and from 1842-44 "in charge of the 3rd Cavalry." From 1844-53 he was Captain Commanding the 4th Cavalry. From 1854-61 in the 2nd Light Cavalry (Madras). He retired as Colonel, December 31st, 1861.

Doria, Richard Andrew.

Born, 1816. Ensign, February 24th, 1835. Lieutenant, April 30th, 1837. Captain, September 3rd, 1846.

Appointed to Hyderabad Contingent, 4th May, 1844.

In 1844 he was in the 1st Infantry, Nizam's Army. In 1845, Acting-Adjutant of the 4th Infantry. 1845-47, Adjutant of the 3rd Infantry. 1847-54, on special duty. Appointed Second in Command of the 3rd Cavalry, February 3rd, 1854. Commandant of the 4th Cavalry, 1854-56. On June 9th, 1856, he vacated the command of the 4th Cavalry.

He served in the operations on the Yangtze Kiang River in 1841-42, and was employed against the Rohillas in 1855, and thanked for his services.

Nightingale, Geoffrey.

Born, September 2nd, 1823. Ensign, June 12th, 1841. Lieutenant, June 24th, 1845. Was appointed to the 10th Native Infantry in 1841.

First appointed to the Nizam's Army, June 4th, 1847, and attached to the 6th Infantry. In 1848 he is shewn as Adjutant, doing duty with the 4th Cavalry. 1849-51, Adjutant of the 3rd Cavalry. 1851-52, Adjutant, in charge of right wing at Ellichpur. On sick leave to Australia, 1852-54. 1854-55, Second in Command of 1st Cavalry. 1855-56, Second in Command of 4th Cavalry. 1857-58, Commandant of the 4th Cavalry. 1859-68, Commandant of the 3rd Cavalry. Promoted Major in 1862; Lieutenant-Colonel in 1868.

War Services.—Served in the campaign in the Hill Tracts of Orissa and Goomsoor, employed with the Hyderabad Cavalry against the Rohillas and Arabs of the Nizam's territory; served in the campaign against Tantia Topee on the bank of the Taptee, and with Brigadier Hill's Field Force in the Nizam's territory, in 1859. Commanded the 3rd Hyderabad Cavalry in the action of Chichambee.

Murray, William.

Born in 1820. Ensign, December 12th, 1840. Lieutenant, January 24th, 1845. First appointed to the Nizam's Army, 24th June, 1851 (his rank in the E.I. Company's Army was Ensign in the 46th Madras N.I.). 1851-53, Adjutant, 1st Infantry, Nizam's Army. 1853, Adjutant, 5th Cavalry. 1854, Second in Command of 2nd Cavalry. 1854-57, Second in Command of 3rd Cavalry. 1857-58, temporarily commanding the 4th Cavalry (with Sir Hugh Rose's force). 1858-70, Commandant of the 1st Cavalry. Promoted Brevet-Major in 1860, and Lieutenant-Colonel in 1867. Proceeded to Europe on sick leave for two years in 1871, and retired in 1873.

War Services.—The campaign in Central India, 1857-58, under Sir Hugh Rose, and present at Piplia, Rawal, Mandesur, Goraria, Madanpur, Tal Bahat, Jhansi, Kotah, Kunch, Golauli, Kalpi, Bilayan, and Gwalior, and wounded by a musket shot in the leg and bayonet wound in the cheek.

Abbott, H. D.

First appears in the Regiment in 1859, as Captain Commanding. Promoted Major in 1859; was Commandant of the Regiment until October, 1868; promoted Lieutenant-Colonel in 1864. Officiated in command of the Hyderabad Contingent, 1865-66. Promoted Colonel, 1867. Commanded the Hyderabad Contingent with the rank of Brigadier-General, 1868-74.

War Services.—Nizam's Dominions, 1849-54. As Brigade-Major in various operations against insurgent Rohillas and Arabs, including the affairs of Woroni and Urni, the reduction of the fort of Rai Man, the siege of Dharur, and the affairs of Aurangabad, Jaswant Pura and Sailur (severely wounded). Indian Mutiny, 1857-58. Campaigns in Malwa and Central India, including the capture of Piplia, the actions of Rawal, Mandesur and Goraria, the forcing of the Dhamoni Pass, the action of Madanpur, the capture of the fort of Tal Bahat, the siege and capture of Jhansi (wounded), the actions of Kunch and Golauli, the advance on Kalpi, the action of Morar (Gwalior), the pursuit of Tantia Topee, the action of Jaora Alipur, and various other affairs (medal and clasp, Brevet-Major, C.B.). Nizam's dominions, 1859, action of Jintur against the Rohillas (in command of a field detachment).

Johnson, A. A.

Officiating Second in Command of the Regiment, 1863-64. Promoted Captain, March 14th, 1865. Commandant of the Regiment from July, 1869, to January, 1884. Promoted Major, March, 1873, and Lieutenant-Colonel, March, 1879.

War Service.—Indian Mutiny, 1857-58. Campaigns in Malwa and Central India, including the capture of Piplia, the actions of Rawal, Mandesur and Goraria, the affair of Baroli, the action of Madanpur, the capture of Tal Bahat, the siege and capture of Jhansi, the action of the Betwa, the affair of Kotah, the actions of Kunch and Golauli, the capture of Kalpi, the action of Morar, the capture of Gwalior, and the pursuit of Tantia Topee. (Medal and clasp).

Nizam's dominions, 1859. Defeat of a body of Rohillas at Chichamba.

Walker, J. G. D.

Attached to the Regiment, June, 1880. Officiating Adjutant, September, 1880 (permanent appointment was squadron officer, 1st Cavalry H.C.). Was subsequently in the 6th Madras Light Cavalry. Commandant of the 4th Cavalry H.C. from April 1st, 1884, to April, 1891. Promoted Lieutenant-Colonel, 1884, and Colonel, 1888.

War Services.—Indian Mutiny, 1859. Operations in Bundelkhund under Brigadier Wheeler. Burma, 1888-89, with 4th Cavalry H.C.

Mason, H. M.

From Squadron Commander, 1st Cavalry H.C. appointed Officiating Commandant, 4th Cavalry, October 10th, 1889. Promoted Major, February, 1890. Commandant of the Regiment, June 6th, 1891, to February 7th, 1902. Promoted Lieutenant-Colonel, February 8th, 1896; Brevet-Colonel, February, 1900. On vacating command appointed Colonel on the Staff. Commanded the Nowshera Brigade. Political Resident and General Officer Commanding Aden from 1904 to 1906. Promoted Major-General. Appointed Hon. Colonel of the Regiment in 1905.

War Service.—Burma, 1886-87, with 3rd Cavalry H.C.

Wapshare, R.

First commission, 1880, Lieutenant, Royal Marines. Joined 4th Cavalry, H.C., from the 3rd Bombay Cavalry as Officiating Squadron Officer, in October, 1884. Officiating Squadron Commander, January, 1886, to January, 1887. Transferred to the 3rd Cavalry, H.C., 1887. Officiating Commandant of the 4th Lancers, H.C., February 9th, 1902. Appointed Commandant of the 4th Lancers, May 18th, 1903. Vacated command, 4th May, 1906. A.A.G. Adjutant General's Division, Army Headquarters, from May, 1906, to March, 1910. Appointed Commandant of the Cavalry School at Saugor, with the rank of Brigadier-General, on April 1st, 1910.

War Service.—Burma, 1886-88, with the 3rd Cavalry, H.C.

Steele, St. C. L.

Educated at Marlborough. First commission, May 11th, 1878, in the 64th North Staffordshire Regiment. Joined 33rd Bengal Native Infantry, March 10th, 1880. Joined 2nd Bengal Lancers, April 2nd, 1881. Adjutant, 2nd Bengal Lancers, 1885-88. Commandant, 30th Lancers, May 5th, 1906, to August 30th, 1909. Appointed A.A.G., 1st Peshawar Division, September 1st, 1909.

Staff Service.—Officiating Inspector Army Signalling, Bengal and Punjab, 1895. D.A.A.G., Pachmari and Allahabad, 1895-1900. A.A.G., 8th Lucknow Division, December, 1905, to May, 1906. Railway Staff Officer and Special Service Officer, China Expeditionary Force, 1900.

War Service.—Egypt, 1882. Action of Kassassin, Battle of Tel-el-Kebir. N.E. Frontier of India, Manipur, served as Brigade Major, Silchar Column. Despatches. N.W. Frontier of India. Waziristan, 1894-95, served as Assistant - Superintendent, Army Signalling, 1st Brigade. Soudan, 1896, Dongola Expedition, Brigade Transport Officer. Tirah, 1897-98, Section Road Commandant. China, 1900, served with the Transport Department, and as Special Service Officer in Manchuria. Mentioned in despatches. Received C.B., 1911, Coronation Honours list.

He was succeeded by Lieutenant-Colonel W. H. Fasken, the present Commandant.

(B) OFFICERS WHO HAVE SERVED IN THE REGIMENT (IN CHRONOLOGICAL ORDER).

Gordon, John William.

Born February 11th, 1804. Educated at Addiscombe College. Appointed Cadet, Bombay Infantry, in 1818, and Ensign, 1820. Lieutenant, August 13th, 1820. On June 28th, 1821, he was appointed Assistant in the Department of Revenue and Topographical Survey of Gujerat, and Assistant Surveyor in the Deccan in 1823.

Appointed Acting-Adjutant to the 1st extra Battalion, 7th Bombay N.I., in 1823, and Officiating Adjutant of the 2nd Battalion in 1824.

He was placed at the disposal of the Resident, Hyderabad, on December 1st, 1827, and served in the 4th Infantry, Nizam's Army; on May 27th, 1831, he was attached as Adjutant to the 4th Cavalry.

General Sleigh reported in 1832, after inspecting the 4th Cavalry, that "It is under the command of Lieutenant Gordon, a young officer, active and intelligent, a capital horseman, and a fit person for the Irregular Horse."

In 1834 he was represented to be particularly fit for the command of the Nizam's 5th Cavalry. The Governor-General, however, objected on the grounds of his comparatively short period of service. (He had only just attained the rank of Captain).

He was nominated to the command of a regiment of the Nizam's Cavalry, November 2nd, 1835. He died in the Neilgherries, July 23rd, 1839. This officer was Officiating Commandant, 4th Cavalry, for a short time.

Malcolm, D. A.

The first record of this officer is in 1838, when he is shewn as a Captain in the 4th Cavalry at Mominabad. In the third quarter of 1838 he was employed in the Thuggee Department; his name disappears from the 4th Cavalry in the last quarter of 1838.

Trower, C. F.

First record is in 1838, when he is shewn as Lieutenant and Adjutant of the 4th Cavalry at Mominabad. Promoted Captain in 1842, and shewn as "withdrawn to Afghanistan" in that year. In 1843 he was Acting-Brigade-Major of the Cavalry Division.

No further record.

Wahab, W. M.

Lieutenant in the 4th Cavalry in 1839. Captain and Officiating Adjutant in 1842; with Regiment as Captain till first quarter, 1847, when he was on leave on medical certificate to the Neilgherries; his name disappears from the Regiment in 1847.

Mayne, H. O.

Lieutenant and Acting-Adjutant in 1842; Adjutant from 1843 to 1849; 1850 he is shewn as Brigade-Major and Paymaster, Ellichpur District. Name subsequently disappears from the Regiment. In 1858 he raised Mayne's Horse, which became subsequently the 1st Central India Horse.

Skinner, H.

Name first appears in Regiment as Captain H. Skinner in 1847, with remark:—"Europe on furlo." In 1848 he was with the Regiment, apparently as Second in Command. In 1849 he is shewn as doing duty with the 2nd Cavalry at Ellichpur. In 1850 he was on leave. Name subsequently disappears from Regiment.

Harrison, C. A.

Name only appears in the Army List once, in 1849, when he is shewn as Lieutenant Harrison. Name disappears in next Army List.

Arthur, E. P.

Acting-Adjutant, 4th Cavalry, in first quarter of 1849. Adjutant in second quarter. Name does not occur again.

Clagett, T. W.

First appears as Captain in 1851; Acting-Adjutant in 1852. In 1853 shewn as commanding wing, 3rd Cavalry at Ellichpur. In 1854, Second in Command of 4th Cavalry at Bolarum. In the third quarter of 1854 he was still Second in Command, Captain Loria being shewn as Commandant. In 1855 name disappears.

O'Connor, E. N. T. R.

Adjutant, 4th Cavalry, in second quarter of 1851. In fourth quarter name disappears; he apparently retired.

Fraser, Hastings.

Shewn as Adjutant, 4th Cavalry, 1853-54. In 1855 Adjutant and Officiating Second in Command—in charge of left wing at Ellichpur. Fourth quarter, 1855, doing duty with 2nd Cavalry. 1856-58 again appears as Adjutant, 4th Cavalry. (In latter year Regiment was with Sir Hugh Rose's force). Name disappears after the third quarter of 1858. He served with 4th Cavalry and as Staff Officer to the H.C. Field Force during the Central India campaign. Was subsequently Military Secretary to Resident, Hyderabad.

Laing, W.C., Surgeon.

In medical charge of the 4th Cavalry from 1838-42.

McLachlan, C., Surgeon.

In medical charge of 4th Cavalry from 1842-1845.

Mackenzie, A. M. W., Surgeon.

In medical charge of the 4th Cavalry from 1846-1855.

Orr, J. H., Surgeon.

In medical charge of 4th Cavalry from 1855-1859; leave to Europe, 1859-60; in medical charge, 1861-62. Served with the 4th Cavalry in Central India; mentioned in despatches, received the C.B.

Burn, A. M., Surgeon.

In medical charge of the 4th Cavalry from 1859-65.

Gib, W. A.

Only appears once, in 1855, as Lieutenant and Acting-Adjutant.

Dowker, H. C., Lieutenant.

Second quarter, 1856, shewn as Adjutant; disappears in next quarter. In 1859 shewn as Second in Command of 4th Cavalry. 1860, Lieutenant Dowker, Second in

Command and in charge of left wing at Ellichpur. 1861, Captain and Second in Command. 1862-64, Major, Commanding the 3rd Cavalry. 1864-65, Second in Command of 4th Cavalry. 1865, Officiating Commandant from 11th May, 1865, to October, 1866.

War Service.—Indian Mutiny, 1857-58. Campaign in Malwa and Central India. Actions of Mandesur and Goraria, forcing of the Dhamoni Pass. Affair of Baroli; action in the Madanpur Pass; capture of Tal Bahat and Chanderi; siege and capture of Jhansi (severely wounded). Affair of Kotah; action of Kunch, capture of Kalpi; action of Bilayan, and capture of Gwalior (several times mentioned in despatches; Brevet of Major). Afghanistan, 1880, with the Kandahar Field Force.

Grant, A., Lieutenant.

1857, Second in Command till 1858. Name disappears from Regiment after 1858.

Dun, E. W.

1858. Lieutenant and Officiating Second in Command. Served in 4th Cavalry with Sir Hugh Rose's force, Central India Campaign. Name disappears after 1858.

Stewart, R. C. (afterwards Sir Richard Stewart, K.C.B.).

First appears in the Army List of 1858 as Adjutant—appointed, not joined. Adjutant, 1859-60. Name disappears from the Regiment in the fourth quarter of 1860.

Appointed Honorary Colonel of the Regiment, May 13th, 1904. Killed at Cheltenham by a fall from his horse when returning from hunting, 1905.

War Services.—India, 1858-59. Served with the Mysore Horse at the action of Shorapore (dangerously wounded); in the pursuit of Tantia Topee, in the valley of the Taptee; and in the affair with the Rohillas at Jintur. Burma, 1887-88. Commanded a Brigade in the Expedition to the Ruby Mines (mentioned in despatches, C.B.).

Burma, 1892-93. Operations in the Chin Hills (K.C.B.).

Prescott, J. C. P.

Shewn as Captain and Second in Command in 1859, apparently left the Regiment in the third quarter of 1859.

Innes, F. J.

1861, Adjutant of the 4th Cavalry at Ellichpur. 1862, Officiating Second in Command. 1865-67, Officiating Second in Command.

War Service.—Indian Mutiny, 1857-58. Storming of Kolhapur, September 10th, 1857. Campaign in the Deccan, 1858. Nizam's dominions, 1859. Defeat of Rohillas at Jintur.

Playfair, A. L.

1862, shewn as Lieutenant, doing duty. 1863, Officiating Adjutant. Apparently disappears from the Regiment in the following year.

Onslow, H. C.

Shewn as Adjutant of the Regiment, 1863-64. Promoted Captain, December 7th, 1867. Name disappears, October, 1873.

War Service.—North Canara, 1858.

Hemans, A. C. W.

(Madras General List, Cavalry). First commission, 20th January, 1862. Appointed to the 4th Cavalry in 1864, "to act as doing duty Officer." January to October, 1866, Officiating doing duty Officer. April to October, 1867, Squadron Subaltern, 1st Cavalry, Officiating Squadron Subaltern, 4th Cavalry.

Briggs, T. C.

Shewn as Lieutenant and doing duty Officer in July, 1865.

Stewart, H. S., Lieut. Madras Staff Corps.

January to October, 1866, Officiating Adjutant (doing duty Officer, 3rd Cavalry).

Promoted Captain, December 4th, 1870. Officiating Second in Command (Adjutant, 3rd Cavalry), January to April, 1876.

War Service.—Afghanistan, 1879-80, with the Kandahar Field Force.

Ottley, R. Madras Staff Corps.

First commission, February 16th, 1860. Officiating Squadron Subaltern, January, 1868. Appointed Squadron Subaltern, April 21st, 1868. Officiating Adjutant, July, 1868, to April, 1869. July, 1869, to January, 1871, shewn as Adjutant, 2nd Cavalry, but Officiating Adjutant, 4th Cavalry. Promoted Captain, December 20th, 1869.

War Service.—Indian Mutiny, 1858-59. Operations in the Saugor and Nerbudda territories.

Hill, E., Colonel. Staff Corps.

Second in Command, January, 1869, to October, 1877. Promoted Lieutenant-Colonel, June 10th, 1874. Officiating A.A.G., Hyderabad Contingent, January to July, 1876. Promoted Colonel, June 10th, 1879. Officiating Commandant, 1st Cavalry H.C., June, 1882, to March, 1883. Officiating Commandant, 4th Cavalry H.C., March to September, 1883. He was Second in Command of the Regiment from 1869 to 1877, with Captain Johnson (afterwards Major and Lieutenant-Colonel) junior to him.

War Services.—Nizam's dominions, 1854-55. Operations against the Rohillas and Arabs. Indian Mutiny, 1858-59. Campaign in Oudh; actions of Nawabgang and Velimpore. Capture of Amethi and Sankarpur, and the final operations on the Nepal Frontier (several times mentioned in despatches).

Talbot, A. C., Lieutenant. Royal Artillery.

Officiating Squadron Subaltern, October, 1869, to January, 1870. In later years Resident, Persian Gulf, then Kashmir.

Powlett, N., Lieutenant. Royal Artillery.

Squadron Subaltern, July, 1870, to April, 1871.
War Service.—Afghanistan, 1878-79.

Gilchrist, R. A., Colonel. Staff Corps.

July, 1871, Squadron Subaltern. With 1st Cavalry H.C. (temporary), October, 1875, to January, 1876. Afterwards commanded 3rd Lancers H.C. and the Madras Brigade.

War Service.—Abyssinia, 1868.

Cummins, J. T., Major-General, C.B., D.S.O. (General List, Madras Infantry).

Joined as Brevet-Captain, October, 1873. Adjutant October, 1873, to October, 1877. Promoted Major, January, 1881. Appointed Second in Command, April 1st, 1874. Lieutenant-Colonel, June 8th, 1887. Officiating Commandant, July, 1887, to January, 1888. Afterwards commanded 2nd Lancers H.C.

War Services.—Afghanistan, 1878-80. Both Bazar Valley expeditions with Khyber Force. Staff Officer, Kuram Valley Transport (Brevet of Major). Egypt, 1882, as Second in Command of Punjab Mule Corps. Soudan, 1885; Battle of Tamai. Burma, 1886-87 (despatches, D.S.O.). Burma, 1888-89, with 4th Cavalry). North China, 1900-01, C.B.

Bird, W. J. B., Lieutenant. Royal Artillery.

Squadron Subaltern, April, 1876, to October, 1877. Officiating Adjutant, March to October, 1876. Squadron Officer, April, 1876, to April, 1879.

Adye, C., Lieutenant. 12th Foot.

Officiating Squadron Subaltern, July, 1876, to January, 1878. Transferred to 1st Cavalry H.C., January, 1878.

War Service.—Afghanistan, 1879-80. Actions of the Shutargardan and Ali Khel; operations at and around Kabul, December, 1879; march from Kabul to Kandahar, and battle of Mazra, September 1st, 1880. Afterwards commanded the 1st Cavalry.

Knox, F. R. B., Lieutenant. Staff Corps.

October 24th, 1876, Officiating Adjutant (Squadron Subaltern, 2nd Cavalry), to April, 1879. June, 1879, Adjutant, Officiating Second in Command, October, 1884, to April, 1885. Promoted Captain, November 13th, 1884. Officiating A.A.G. H.C., October, 1886, to October, 1887. Transferred to 3rd Cavalry H.C., October, 1888. Attached as Officiating Second in Command, October, 1890, to April, 1891. Second in Command from June, 1891, to April, 1893, when he retired from the Service.

War Service.—Burma, 1887-89, with 4th Cavalry (despatches).

Fitzgerald, C. J. O., Colonel. Madras Staff Corps.

Officiating Commandant from March 9th, 1877, to January, 1879. (Permanent appointment was Squadron Commander, 3rd Cavalry H.C.). Afterwards commanded the 3rd Cavalry H.C.

War Services.—India, 1857-59; Afghanistan, with Kandahar Field Force; Burma, 1886-88 (in command of the 3rd Cavalry, mentioned in despatches, awarded the C.B.), K.C.B. in 1907.

Meade, J. W. B., Lieutenant. Royal Artillery.

Officiating Adjutant, March, 1881. Officiating Squadron Officer, 1882-83. Transferred to 3rd Cavalry H.C., which he subsequently commanded.

War Services.—Afghanistan, 1878-80, actions of Ahmed Khel and Arzai. Burma, 1886, with 3rd Cavalry.

Johnstone, R. F. M.

Lieutenant, 16th Foot. Officiating Squadron Officer, 1881. Adjutant, 1884-89. Promoted Captain, May 11th, 1889. Second in Command, April, 1893. Officiating Commandant, April, 1896, to April, 1897. Promoted Major, May 11th, 1898. Retired from the Service, October, 1898.

War Services.—Burma, 1888-89, with 4th Cavalry. South Africa, as a District Commandant; operations in Cape Colony, May, 1901-1902.

Gubbins, C. E.

Captain and Officiating Second in Command, September, 1882, to January, 1883. Transferred to the 3rd Cavalry H.C.

War Service.—Burma, 1886-88, with 3rd Cavalry.

Mason, S. M.

Lieutenant, Bombay Staff Corps. Officiating Squadron Officer, 4th Cavalry, June 29th, 1887. Appointed Squadron Officer, November, 1888. Adjutant, January 22nd, 1890. Promoted Captain, October 22nd, 1892. Appointed Squadron Commander, April 11th, 1893. Second in Command, February 13th, 1899. S.S.O. 1st Class, Bolarum, July, 1897, to February, 1899. Died, December 22nd, 1899, when on leave to England.

War Service.—Burma, 1886-89. Operations of the 6th Brigade under Brigadier-General R. C. Low. Special operations in the Sagaing District under Colonel W. P. Symons (mentioned in despatches). Served with 4th Cavalry in the Burmese War, 1888-89.

Leslie, T. D.

Lieutenant, Madras Staff Corps. Officiating Squadron Officer, January 16th, 1888. Transferred to 2nd Cavalry H.C. Promoted Captain, September 30th, 1894. Appointed to the 3rd Madras Lancers from the half-pay list. Officiating Squadron Commander, 4th Lancers H.C., August 10th, 1894, to June, 1895. Transferred to 2nd Lancers H.C., Officiating Commandant, 3rd Lancers H.C., from 1901 to April 30th, 1903. Promoted Major, September 30th, 1903. Second in Command, 4th Lancers H.C., from May 1st, 1903. Died at Poona, August 10th, 1904.

War Service.—Burma, 1888-89, attached to 4th Cavalry.

Beddoes, H. R.

Lieutenant, 7th Hussars. Officiating Squadron Officer, January 25th, 1888, to October, 1889, when he reverted to British Service.

War Service.—Burma, 1888-89, attached to 4th Cavalry.

Subsequently, in 1900 and 1902, he saw a good deal of service in West Africa, being severely wounded, mentioned in despatches, and receiving a Brevet.

McSwiney, E. F. H., Brigadier-General, C.B., D.S.O.

Transferred from 3rd Cavalry H.C., as Squadron Officer, August 2nd, 1888. Adjutant, May, 1889, to January, 1890. Officiating Second in Command, April, 1890. Promoted Captain, January 22nd, 1890. Paid Attaché, I.B.Q.M.G.'s Department, January, 1892. D.A.Q.M.G., October, 1892. Transferred to 1st Lancers H.C. as Second in Command, July, 1895. Subsequently commanded 1st Lancers H.C. and the Ambala Cavalry Brigade. Died while holding the latter command.

War Services.—Afghanistan, 1880. Burma, 1886-88 (despatches, D.S.O.). N.W. Frontier of India, 1897-98. D.A.A.G., Kurram-Kohat Force. Action of the Ublan Pass; operations on the Samana and in the Kuram Valley. Relief of Gulistan (despatches). Tirah, 1897-98. D.A.A.G., Kurram Moveable Column (despatches, Brevet of Lieutenant-Colonel).

Wyllie, F.

Madras Staff Corps. Appointed Squadron Officer, 1st Cavalry. Subsequently served in 2nd Lancers H.C. Transferred as Second in Command, 4th Lancers H.C., 11th July, 1900. Transferred back to 2nd Lancers H.C., as Second in Command, February 7th, 1902.

War Services.—South Africa, 1879 Zulu War. Burma, 1888-89, attached to 4th Cavalry.

Nelson, F. J.

Madras Staff Corps. Officiating Squadron Officer, June 12th, 1889. Appointed Squadron Officer, April 1st, 1891. Burma Military Police, January, 1892, to February, 1893. A.D.C. to H.H. the L. G. of N.W.P. and Oudh, April, 1894, to January, 1896. Promoted Captain, February 28th, 1896. Appointed Squadron Commander, May 14th, 1897. Died of cholera at Hingoli, August, 1897.

War Services.—Bechuanaland, 1884-85. Burma, 1886-87 (mentioned in despatches). Burma, 1892, commanded the Sinkan Column against the Kachins, action of Tonhon (severely wounded).

Davidson, A. C.

Transferred from 2nd Cavalry H.C., as Officiating Squadron Officer, June 12th, 1889. Appointed Squadron Officer, July, 1890. Officiating Squadron Commander, July, 1891. Transferred, October, 1891, to H.C. Infantry, and died shortly afterwards.

Keille, Surgeon-Major. I.M.S.

Served for many years as Medical Officer, 4th Cavalry. Transferred to 2nd Lancers H.C., December 1st, 1894.

War Services.—Burma, 1888-89, with 4th Cavalry. Tirah, 1897-98.

Jones, E. G.

Second Lieutenant, S.W. Borderers. Officiating Squadron Officer, September 16th, 1891, to January, 1893, when he was permanently transferred to Burma Infantry.

Henderson, M. H.

Second Lieutenant, 7th Dragoon Guards. Officiating Squadron Officer, November 8th, 1892. Squadron Officer, July, 1893. Adjutant (sub. *pro tem.*), September, 1894, to October, 1895. Officiating Squadron Commander, January, 1896, to January, 1897. Adjutant, April, 1897, to 1901. Transferred as 4th Squadron Commander to the 26th Light Cavalry, June, 1904.

Goldthorp, F. H.

Lieutenant, Norfolk Regiment. Officiating Squadron Officer, April, 1893, to January, 1894, when he was transferred to Punjab Cavalry.

War Services.—N.W. Frontier of India, 1894-95, Waziristan. N.W. Frontier of India, 1897-98, operations on the Samana, Kurram. China, 1900-01.

Beville, H. G. P.

Lieutenant, Suffolk Regiment. Joined from 2nd Bombay Lancers. Officiating Squadron Officer, July 16th, 1893, to October, 1894. Subsequently joined the S. and T. Corps.

Bentinck, R. J. Educated Wellington College and R.M.C., Sandhurst.

First Commission, 5th February, 1887, 1st West India Regiment. Joined 4th Bombay Rifles, 5th February, 1889. 5th Bombay Light Infantry, December, 1890. 1st Lancers H.C., November, 1891. 4th Lancers H.C., October, 1892. Retired from the Service, 5th February, 1902.

Staff Service.—A.D.C. to Governor, West Africa, January—July, 1888. Personal Assistant to Resident, Hyderabad, 1891-92. A.D.C. to H.E. The Viceroy of India, January, 1893—December, 1893.

War Services.—South Africa, November, 1899, to June, 1902, with Thorneycroft's Mounted Infantry and 1st Imperial Light Horse.

Sargent, Brigade-Surgeon Lieutenant-Colonel. I.M.S.

Joined from 2nd Lancers H.C., November 17th, 1894. Died at Hingoli, July 20th, 1895.

Wortabet, H., Surgeon-Major. I.M.S.

Appointed Medical Officer, November 1st, 1895. Retired as Colonel, June 22nd, 1909.

War Services.—Afghanistan. Burma, 1886-88, with 3rd Infantry H.C.

Dalgliesh, E. E. C.

Lieutenant, Dorset Regiment. Officiating Squadron Officer, March 5th, 1896, to October, 1898. Transferred to 1st Lancers H.C., October 21st, 1898.

Goodfellow, R. C.

Officiating Squadron Officer, September 10th, 1896, to April, 1900. Transferred to the Central India Horse.

Armstrong, A. K.

Captain (Squadron Officer, 1st Lancers H.C.)., Squadron Commander, *pro tem.*, 4th Lancers, October, 1897, to January, 1898.

War Service.—South Africa, 1899-1900, with Mounted Infantry.

Humfrey, B. J. H.

Squadron Commander, *pro tem.*, January 3rd, 1898, to January, 1899. Joined from the 3rd Madras Lancers. Transferred to 1st Lancers H.C., February 15th, 1899.

Newnham, P. F., Lieutenant.

Transferred from 3rd Lancers H.C. as Squadron Officer, January 2nd, 1899. Served with Thorneycroft's Horse in the South African War; killed in action at Spion Kop, January 24th, 1900.

Hicks, P. S., Lieutenant.

Originally appointed to 1st Lancers H.C., then transferred to 3rd Lancers H.C. Transferred from the 3rd Lancers, April 1st, 1900, while on furlough. Served in the 2nd provisional Hussars in England, 1900-01. Retired from the Service, 1901, without returning to India.

Percy-Smith, V., Lieutenant.

Squadron Officer, July 3rd, 1900. Transferred to the 1st Lancers H.C., January 31st, 1901. Shortly after resigned his commission.

Halliday, S. C., Lieutenant.

Squadron Officer, July 30th, 1900, to 1903. Promoted Captain, October 21st, 1902. Resigned his commission, 1903.

Rawlins, C. W., Major.

Joined from the 12th Cavalry as Second in Command, October 16th, 1904. Appointed Assistant Military Secretary to L.G.C., Western Command (Sir Archibald Hunter), May, 1905. Died of enteric at Poona, September 19th, 1906.

War Services.—Egypt, 1882. Chitral Expedition, 1895, relief of Chitral as Assistant Superintendent of Army Signalling. N.W. Frontier of India, 1897-98. Operations in the Tochi Valley, as Superintendent of Army Signalling (despatches). East Africa, Somaliland, served with Mounted Infantry.

Warner, W. W., Major.

First Commission, 1888. Joined 3rd Cavalry H.C. from the Poona Horse. Adjutant, 3rd Cavalry. Appointed to the 4th Lancers H.C., April 1st, 1903. Inspecting Officer, Mysore and Hyderabad I.S. Troops, 1906 to 1907. Retired from the Service, December 21st, 1907.

Turner, J., Lieutenant.

Joined from the unattached list, 1906. Transferred to the 116th Mahrattas, July 26th, 1909.

Trelawny, F. C., Lieutenant.

Joined from the Suffolk Regiment, November, 1906. Transferred to the 69th Punjabis, April 30th, 1908.

(C) OFFICERS SERVING IN THE REGIMENT ON JANUARY 1st, 1911.

Mason, H. M., Major-General.

Appointed Colonel, 1905. *See past Commandants.*

Fasken, W. H., Lieutenant-Colonel.

Born, May 9th, 1863. Educated, Marlborough and Sandhurst. First commission, May 10th, 1882, in 2nd Battalion Lincolnshire Regiment. Joined Indian Staff Corps, 10th Bengal Lancers, 15th October, 1885. Adjutant, 1889-92. Appointed Commandant, 30th Lancers, 1st September, 1909.

Staff Service.—Brigade-Major to I.G. of Cavalry, October, 1893, to May, 1894. Inspecting Officer, I.S. Cavalry (Mysore and Hyderabad), June, 1895, to December, 1900. Officiating A.A.G., 3rd Division, April to November, 1908.

Elliot, C. R., Major.

Born, 10th October, 1865. Educated, Sandhurst. First commission, 30th January, 1886, Oxfordshire Light Infantry. Transferred to Middlesex Regiment, 17th February, 1886. Joined 3rd Infantry, Hyderabad Contingent, 31st March, 1887. Joined 3rd Cavalry, Hyderabad Contingent, 1st April, 1888. Adjutant of 3rd Cavalry, 1st February, 1891—1st May, 1895. Transferred to 4th Lancers, Hyderabad Contingent, 24th September, 1897. Second in Command, 17th September, 1906.

War Service.—Burma, March, 1887—March, 1888.

Stotherd, E. A. W., Major.

Born, 26th March, 1864. Educated, Marlborough and Sandhurst. First commission, 5th February, 1887, 2nd West India Regiment. Joined Bengal Staff Corps, 1st April, 1888. Joined 4th Cavalry H.C., 18th November, 1888.

Special Services.—Reconnaissance in Persia, 1893. Recommended MacGregor Memorial Medal.

Staff Services.—Staff Lieutenant, I.B., Q.M.G.'s Department, 1895. Attaché, A.G.'s Division, 1900. Attached to General Staff, Army Headquarters, June, 1910—May, 1911. Officiating General Staff Officer. Wrote the Military Report on Persia.

War Services.—Burma, 1888-89, with 4th Cavalry. N.W. Frontier of India, 1897-98; operations on the Samana, September 1897. Tirah: Actions of Dargai and of the Sampagha and Arhanga Passes. Operations in the Waran Valley and against the Khani Khel Chamkannis. Operations in the Bara Valley, 7th—14th December, 1897. North China, 1900-01, relief of the Legations at Peking, and expedition to the Western Hills and Pa-ta-chou.

Lane, H. A., Major.

Born, 6th October, 1868. Educated, Sandhurst. First commission, 14th September, 1887, 3rd Dragoon Guards. Attached 4th Infantry H.C. Joined 2nd Cavalry H.C., January 7th, 1889. Joined 4th Lancers H.C., August 8th, 1899.

Staff Services. — Secretary to the Resident, Hyderabad, September 18th, 1894, to April 2nd, 1898. Personal Assistant to ditto, April 3rd, 1898, to August 7th, 1899. Commandant, 57th Silladar Camel Corps, March, 1903, to April, 1908.

War Services.—East Africa, Somaliland. Despatches, "London Gazette," 9th September, 1904.

Barnard, A. E., Major.

Born, 30th January, 1867. Educated, Oxford Military College and Sandhurst. First commission, 11th February, 1888. 2nd Battalion Derbyshire Regiment. Joined 4th Lancers H.C., October 15th, 1891. Burma Military Police, December, 1895, to January, 1901. Commandant, Silladar Camel Corps, January, 1908, to January, 1910.

War Services.—N.W. Frontier of India, 1st Miranzai Expedition, 1891. East Africa, 1902-04, operations in Somaliland. N.W. Frontier of India, 1908, operations in the Zakka Khel country; operations in the Mohmand country.

Leslie, P. N., Major.

Born, August 14th, 1868. Educated, Sherborne School and Sandhurst. First commission, November 29th, 1890, Derbyshire Regiment. Joined 5th Infantry H.C., May, 1893. Joined 4th Lancers H.C., January 5th, 1894. Burma Military Police from July, 1899, to September, 1904. Appointed to Cantonment Magistrates' Department, 1907.

Harbord, C. R., Major.

Born, December 2nd, 1873. Educated, Sandhurst. First commission, 3rd September, 1892. I.S.C., unattached list. Attached Northamptonshire Regiment, December 25th, 1892. Joined 24th Madras Infantry, December 26th, 1893; 2nd Infantry H.C., 27th January, 1894; 5th Infantry H.C., 29th May, 1894; 1st Lancers H.C., 22nd January, 1895; 2nd Lancers H.C., 2nd July, 1897; 30th Lancers, 21st October, 1908.

Staff Services.—Officiating D.A.A.G., Secunderabad, 29th April, 1908, to 20th October, 1908. Staff Officer to the Inspector of Cavalry from December, 1911.

War Services.—South African War, 1899-02. As special service officer. Operations in the Transvaal, west of Pretoria, July to November 29th, 1900, including actions at Venterskroon, 7th—9th August. Operations in the Transvaal between 30th November, 1900, and May 31st, 1902. Operations in Cape Colony between 30th November, 1900, and 31st May, 1902. East Africa, 1902, operations in Somaliland.

Muspratt, V. E., Captain.

Born, June 28th, 1875. Educated, Sandhurst. First commission, September 28th, 1895. Royal West Kent Regiment. Joined 3rd Lancers H.C., October 7th, 1899. Adjutant, 3rd Lancers, October, 1901, to April, 1903. Transferred to 4th Lancers H.C., April 1st, 1903. Adjutant, 4th Lancers, September, 1903, to November, 1905. Passed Staff College, Quetta, 1907-08. Officiating General Staff Officer, Army Headquarters, June 20th, 1910.

War Service.—N.W. Frontier of India, 1897-98. Malakand, Buner. Actions at Landikai and the Tanga Pass.

Wilford, E. E., Captain.

Born, January 13th, 1876. Educated, Sandhurst. First commission, 5th September, 1896, East Yorkshire Regiment. Joined 5th Madras Infantry, 19th December, 1898; 1st Bengal Infantry, 7th January, 1900; 5th Punjab Infantry, 5th March, 1901; 4th Lancers H.C., 20th July, 1901. Burma Military Police, April, 1906, to April, 1911.

Fellows, B. C., Captain.

Born, 1st August, 1877. Educated, Sandhurst. First Commission, August 4th, 1897, I.S.C., unattached list. Attached Cheshire Regiment, 22nd November, 1897. Joined 2nd Lancers H.C., 1st December, 1898. Joined 4th Lancers H.C., 6th March, 1899. Adjutant, 23rd September, 1901, to 31st July, 1903.

Staff Services.—A.D.C. to Political Resident, Aden, July, 1904, to August, 1906. A.D.C. to Governor of Bombay, September, 1906, to October, 1907. Attaché, Army Headquarters, A.G.'s Division, March to October, 1909.

Gledstanes, A. U., Captain.

Born, July 6th, 1876. Educated, Harrow. First commission, December 1st, 1897, The Royal Scots. Joined 39th Garhwal Rifles, 4th July, 1901. Joined 4th Lancers H.C., 19th September, 1902. Adjutant, 1st November, 1905, to 31st October, 1909.

Walker, H. T., Captain.

Born, 8th January, 1879. Entered Army through Militia. First commission, May 20th, 1899, the Connaught Rangers. Joined 1st Lancers H.C., 19th January, 1903. Joined 30th Lancers, 13th July, 1903. Appointed to Burma Military Police, April, 1911.

Raymond, E. D., Captain.

Born, May 2nd, 1881. Entered Army through the Militia. First commission, April 18th, 1900, Essex Regiment. Joined 30th Lancers, 12th March, 1907.

War Service.—South African War, 1900-02. Operations in the Transvaal, east of Pretoria, including the actions at Belfort, 26th and 27th August. Operations in the Transvaal west of Pretoria, including actions at Frederickstaad, 17th—25th October. Operations in the Transvaal and Cape Colony, 30th November, 1900, to January, 1902.

Maydwell, H. S. L., Captain.

Born, February 8th, 1881. Entered Army through the Militia. First commission, May 5th, 1901, Royal Munster Fusiliers. Attached Connaught Rangers, 27th March, 1902, to 19th September, 1903. Joined 4th Lancers H.C., 19th September, 1903.

War Service.—South African War, 1899-1901. Operations in the Transvaal, February — July, 1901. Operations in Cape Colony, July—October, 1901.

Morris, G. P., Captain.

Born, March 12th, 1882. Educated, Sandhurst. First commission, May 8th, 1901, I.S.C., unattached list. Attached Suffolk Regiment, 7th November, 1901. Joined 3rd Infantry H.C., 12th November, 1902. Joined 30th Lancers, 25th June, 1904. Adjutant, 1st November, 1909.

Staff Service.—Attaché, Chief of Staff's Division, July to December, 1907. Attaché, A.G.'s Division, December, 1907, to February, 1908.

Butler, R. B., Lieutenant.

Born, January 26th, 1884. Educated, Trinity College, Dublin. Entered Army through the Militia. First commission, August 5th, 1905, I.A., unattached list. Attached Royal Irish Fusiliers, 14th December, 1906. Joined 30th Lancers, 15th December, 1907.

Lucas, H. de N., Lieutenant.

Born, December 6th, 1886. Educated, Wellington and Sandhurst. First commission, August 5th, 1905, I.A., unattached list. Attached Royal Scots, 19th October, 1905. Joined 30th Lancers, 6th September, 1907.

Odlum, W. H., Captain. I.M.S.

Born, August 2nd, 1873. Educated, Trinity College, Dublin. First commission, November 14th, 1900, R.A.M.C. Transferred to I.M.S., 19th December, 1907. Appointed to Central India Horse. Transferred to 30th Lancers, 1st September, 1909. Specially noted for medical work performed in India during 1903.

War Service.—South African War, 1900-02. Served as a civil surgeon. Slightly wounded. Operations in Orange River Colony (May to 29th November, 1900), including actions at Biddulphsberg, Bethelhem (6th—7th July), Wittebergen (1st—29th July), Witpoort, Ladybrand (2nd—5th September), Caledon River (27th—29th November). Operations in the Orange River Colony, 30th November, 1900, to 31st May, 1902. Despatches, "London Gazette," 15th November, 1901.

INDEX

ABBOTT, Brig.-Gen., 35, 45, 47, 50, 51, 68, 69, 71, 74, 77, 83, 104, 110, 122, 132, 135-7, 141, 142, 182.
Abdul Karim Khan, Ris. Major, 58, 59, 61, 63.
Abdul Majid Khan, Ris. Major, 56, 58
Adye, Goodson, 191
Ahmad Baksh Khan, 35
Ajunta, 12, 37
Amir Muhd. Khan, Jem., 54
Arabs, 42, 142
Arms, 14, 23, 28, 30, 31, 32
Armstrong, A. K., 196
Arthur, E. P., 186
Atma Singh, 58
Aurungabad Division, 12

BAGNOLD, Brigadier, 43
Bahadurpur action, 128, 129
Banda, Nawab of, 47, 105, 112, 123
Banpur, Raja of, 80, 82, 85, 108
Barnard, A. E., 61, 199
Barwa, Sagar Fort, 85, 91
Beatson, Brigadier, 43, 45
Beaufort, Duke of, 59
Beddoes, H. R., 154, 155, 159, 161, 162, 163, 193
Blair, Col. James, 41, 42
Bell, Col., 52
Belura, 56, 57
Bentinck, R. J., 195
Berar, 1, 32, 48
Betwa, battle of the, 91-98
Beville, H. G. P., 195
Bhils, 12, 36, 37
Bilayan action, 124-5, 143
Bird, W. J. B., 191
Briggs, T. C., 189
Burmese Campaign—
  British officers, 3rd Cavalry, 148
  Strength, 3rd Cavalry, 154
  Casualties, 3rd Cavalry, 154
  British officers, 4th Cavalry, 163-4
  Strength, 4th Cavalry, 164
  Casualties, 4th Cavalry, 164
  Kumri, 164
  Honours and rewards, 166
Burn, Surgeon A. M., 187
Burton, R. G., 25, 34, 144-5, 147-8, 165
Butler, R. B., 202
Byam, Capt., 39, 40, 179

CENTRAL India Campaign—
  Cavalry camps at Jhansi, 88-90
  Cavalry officers who took part, 140
  Casualties, officers, 141
  Casualties, men, 143
  Honours and rewards, 141-2
  Gen. Order C. in C., 127, 128
  Gen. Order Resident, 138, 139
Chanderi Fort, 82, 84, 85
Chichambah action, 48
Chotay Khan, Jem., 122
Clagett, Capt., 45, 186

Clerk, Lieut., 71, 72, 93, 95
Clerk's risala, 17
Clogstoun, 35
Coronation, London 1911, 63, 64
Crossthwaite, Sir Cahs., 154
Cummins, Major-Gen. J. T., 52, 53, 154, 156, 161, 162, 166, 191

DALGLIESH, E. E. C., 196
Daniel, Capt., 47
Davidson, A. G., 194
Davidson, Capt. 41, 137, 180
Davies, Capt. Evan, 7, 13-15, 23, 34
Deccan Horse, 20th, 17, 33
Deccan Horse, 29th Lancers, 17, 33
Delhi Durbars, 59, 64
Deopura actions, 115-117
Dhabee Fort, 37
Dhamani Pass, 81, 83
Dhar action, 66, 143
Dharoor Fort, 44
Doria, Capt., 46, 47, 181
Dowker, Lieut., 35, 50, 85, 102, 104, 111, 122, 125, 187-8.
Drew, 5
Dun, Lieut., 103, 104, 111, 125, 188
Durand, Col. H. M., 66, 72, 77

EDLABAD, 67
Ellichpur Brigade, 13, 24
Ellichpur Horse (5th Cavalry), 10, 16, 18, 21, 22
Elliot, C. R., 198

FASKEN, Lieut.-Col. W. H., 63, 198
Fellows, B. C., 200-201
Firoz Shah, 66, 73
Fitzgerald, Col. C. J. O., 51, 148, 149, 150, 153, 166, 192
FitzPatrick, Sir Denis, 53
French, Col. John, 54
GALL, Major, 67, 72, 74, 88, 98, 106, 107, 109, 121

Garhakota, 81
Garstin, Capt., 41, 180
Gaselee, Lt.-Gen. Sir A., 58
Gib, W. A., 187
Gilchrist, Col. R. A., 190
Gledstanes, A. U., 201
Golauli, 115, 117
Goldthorp, F. H., 195
Goodfellow, R. C., 196
Goraria, 75, 76
Gordon, Capt. J. W., 185
Gordon Highlanders, 175-6
Gordon, Sir John, 8, 16, 17, 36, 38, 173-6, 179
Gough, Lord, opinion H.C. Cavalry, 25
Govind Rao, Duffadar, 56, 58
Grant, A., 188
Grover, Maj.-Gen., 62
Gubbins, C. E., 148, 149, 151, 166, 192
Gumsur, 39, 40, 41
Gurmukh Singh, Kote Duf., 62
Gwalior operations, 128-135

HAIG, Major-Gen. D., 60
Halliday, S. G., 197
Hallis, Capt., 17
Hamilton, Sir Robert, 77
Hampton, George, 34, 46
Harbord, C. R., 64, 200
Hare, Capt., 46, 67, 79, 81, 103, 104
Harrison, G. A., 186
Hastings Fraser, Lieut., 46, 47, 67, 69, 72, 77, 95, 98, 111
Hemans, A. G. W., 189
Henderson, M. H., 195
Hicks, P. S., 196
Hill, Brigadier, 48, 50
Hill, Col. E., 190
Hingoli Brigade, 13, 24
Horses, 20, 23, 27, 31, 165
Hounds, Regimental packs, 59
Humfrey, B. J. H., 196
Hyderabad Division, 12

INNES, F. J., 189

JALNA, 29, 51
Jaora-Alipur action, 135
Jhansi siege, 85, *et seq.*
Jhansi, storming of, 98-105, 143
,, Rewards for conspicuous gallantry, 105
Johnson, Lieut.-Col. A. A., 51, 71, 77, 111, 183
Johnstone, R. F. M., 154, 156, 161, 163, 192
Jones, E. G., 195

KALPI Operations, 113-123, 143
Kellie, Surgeon-Major G., 55, 154, 161, 163, 194
Kesar Singh, Duffadar, 52
Knox, F. R. B., 154, 157, 158, 159, 162, 163, 191
Kotah-ki-Serai action, 133
Kotra action, 108
Kunch battle, 107-11, 143
Kurnool, 41

LAING, Surgeon W. C., 187
Lane, H. A., 199
Leslie, P. N., 199
Leslie, T. D., 59, 60, 154, 159, 161, 193
Lohari, 107
Lucas, H. de N., 202
Luck, Major-Gen. Geo., 53, 54
Lumsden, Brigadier, 50, 51

MACINTYRE, Capt. 45
Mackenzie, Surgeon, 142, 187
Macquoid, Lieut., 125
Madanpur Pass action, 81, 83, 84
Malcolm, D. A., 185
Mandesur action, 66, 73, 75, 77, 143
Mason, Major-Gen. H. M., 53, 58, 60, 148, 183, 198
Mason, S. M., 57, 154, 155. 159, 160, 162, 192-3
Maydwell, H. S. L., 201
Mayne, Brigadier, 45, 46
Mayne, H. O., 186
Maxwell, Col., 114, 115, 117, 121
McLachlan, Surgeon, 187
McSwiney, E. F. H., 148, 150, 151, 152, 166, 193
McVittie, Surgeon-Major, 148
Meade, J. W. B., 148, 192
Mehidpur, 9
Mehidpur (Central India), 70
Mir Mahbub Ali, 39
Mir Muzaffar Ali, Ressaidar, 157
Mirza Wazir Ali Beg, 39
Mirza Zulfikar Ali Beg, R.M., 72
Mominabad Cavalry Brigade, 13, 24
Morar action, 132
Morris, G. P., 202
Muhd. Ali Khan, Duffadar, 52
Muhd. Amir Khan, Duffadar, 55, 57
Muhd. Fazl Khan, Duffadar, 158-9
Muhd. Muzaffar Khan, R.M., 52, 53, 158, 166
Muhd. Oomer Khan, R.M., 50, 142
Mules, 31
Murray, Capt., 47, 67, 68, 69, 71, 72, 74, 77, 104, 111, 116, 142, 182
Murtaza Yar Jung, 16, 34
Muspratt, V. E., 59, 200
Mysore Silladar Horse, 7

NAIKS, 2, 3, 9, 36, 37
Nanpara Cavalry Cup, 61, 62, 63
Namdar Khan, Nawab, 13
Napier, Brigadier, 135
,, despatch, 136
Narut Pass, 81, 82
Nasib Ali, Trooper, 157
Natha Singh, Jemadar, 59, 64
Nelson, F. J., 56, 194
Newnham, P. F., 57, 196

Nightingale, Capt., 47, 68, 181
Nixon, Major-Gen., J. E., 61
North China, 58
Nowah, 9

O'CONNOR, E. N., 187
Odlum, W. H., 202
Onslow, H. C., 189
Order of St. John of Jerusalem, 62
Orr, Capt. S. G., 66, 76
Orr, Lieut.-Col. W. A., 49, 67, 70, 77, 79, 82, 83, 93, 104, 108, 124-6, 131, 141, 142
Orr, Surgeon, 69, 71, 77, 111, 141, 187
Ottley, R., 190

PAY, 4, 5, 18, 19, 21, 22, 23, 26, 29, 30, 32
Pedler, Capt., 14
Percy-Smith, V., 197
Pindarees, 2, 3, 9, 36
Piplia Fort, 68, 69
Playfair, A. L., 189
Polo Cups, 61
Ponies, 31
Powlett, N., 190
Prescott, J. C. P., 189
Punch, 106
Punishment of failures, 38
Punniar, 131

RAHATGARH Fort, 80
Rai Mhow, 44
Rani of Jhansi, 87, 101, 102, 111, 123, 128, 133
Rao Sahib, 47, 105
Rawal action, 71-72
Rawlins, Major G., 60, 197
Raymond, E. D., 201
Reformed Horse, 7, 8, 9, 10
Reorganisation, 1816, 7
" 1826, 11
" 1854, 24
" 1875, 26
" 1902, 32
Report on Nizam's Cavalry, 1832, 20
Retribution Hill, 101, 104
Riddell, Col., 130, 131, 135
Robertson, Lieut.-Col., 123, 129
Rohillas, 36, 39, 41, 42, 43, 44, 48, 49, 102, 142
Rose, Lieut., 134
Rose, Sir Hugh, 35, 47, 77, 80, 82, 85, 91, 95, 96, 105, 109, 116, 119, 120, 131
Rose, Sir Hugh, Despatches, Orders, etc., 86, 95, 103, 118, 122, 126, 130, 137-8
Royal Humane Society Medals, 62
Rur Singh, Jemadar, 63
Russell Brigade, 6, 18
Russell Cavalry, 9, 10
Russell, Resident, 4, 6

SALABAT Khan, 5, 7, 10, 11, 18
Samwell, F., 68, 69, 71, 72
Sanderson, Assist. Surg., 71, 111, 125
Sanoda Fort, 81
Sanoo Khan, Trooper, 161
Sargent, Bde. Surg. Lt.-Col., 55, 196
Saugor, relief of, 80, 81
Sayyid Nur Ali, Jemadar, 98
Shahgarh, Raja of, 83, 108
Shah Mirza Beg, R.M., 50
Shaik Muhd. Sultan, R.M., 54, 55
Sikander Ali Beg., Resdr. 98
Sinclair, Capt., 68, 103
Skinner, H., 186
Sleigh, Major-Gen., 20
Smith, Brigadier, 130, 131—134
Smith, H. B., 8, 10, 17
Soobhan Khan, 5
Soobhan Khan, Jemadar, 160-1
Standards, Regimental, 171, 172

Steele, Col. St. G., 61, 63, 184
Stewart, Brigadier Chas., 78
Stewart, Lieut.-Gen. Sir R., 59, 60, 148, 166, 188
Stewart, H. S., 189
Stotherd, E. A. W., 54, 63, 162, 163, 198-9
Strange, Capt., 43, 181
Stuart, Brigadier C. S., 66, 78, 94, 129, 131, 136
Sutherland, Capt. J., 10, 34, 37
Sydenham, Lieut., 5, 13

TAL Bahat Fort, 85
Talbot, A. C., 190
Tantia Topee, 47 48, 91, 96, 98, 111, 123, 135
Taylor, Brig.-Gen., Report, 40
Tehri action, 117
Transport animals, 31
Trelawney, F. C., 197
Trower, C. F., 186
Turner, J., 197

UNIFORM, 167-172

WAHAB, W.M., 186
Walker, Col., 51, 53, 154, 157, 158, 159, 161, 163, 183
Walker, H. T., 201
Waller, Lieut., 134
Wapshare, Brig.-Gen. R., 58, 59, 61, 148, 150, 152, 184
Warner, W. W., 59, 197
Wells, Lieut., 8, 17
Westmacott, Lieut., 111, 125
White, Sir George, 146
Whitlock, Maj.-Gen., 78
Wilayet Ali Khan, R.M., 53 54
Wilford, E. E., 200
Wolseley, Brig.-Gen., 161, 163
Wortabet, Col. H., 196
Wright, Brig.-Gen. T., 26, 27 29, 51
Wright, E. L., 148, 150
Wyllie, F., 154, 157, 162, 194

YE-U, 149, 156, 162

ADDENDA

www.ingramcontent.com/pod-product-compliance
Ingram Content Group UK Ltd.
Pitfield, Milton Keynes, MK11 3LW, UK
UKHW041848190726
13854UKWH00002B/774

9 781845 743208